DIFFERENT ENGINES

how science drives fiction and fiction drives science

Mark L. Brake and Neil Hook

Macmillan

London New York Melbourne Hong Kong

First published 2008 by
Macmillan
Houndmills, Basingstoke, Hampshire RG21 6XS and
175 Fifth Avenue, New York, N. Y. 10010
Companies and representatives throughout the world

ISBN-13: 978–0–230–01980–5
ISBN-10: 0–230–01980–3

This book is printed on paper suitable for recycling and made from fully managed and sustained forest sources. Logging, pupling and manufacturing processes are expected to conform to the environmental regulations of the country of origin.

A catalogue record for this book is available from the British Library.

A catalog record for this book is available from the Library of Congress.

10   9   8   7   6   5   4   3   2   1
17   16   15   14   13   12   11   10   09   08

Printed and bound in China

CONTENTS

ACKNOWLEDGEMENTS

A version of Chapter 1 appeared as 'Science, fiction and age of discovery', by Mark Brake and Neil Hook, in *Physics Education*, **42**(3), May 2007, pp. 245–52.

A taster entitled 'Copernicus and the wild goose chase' appeared in both *Viewpoint*, the newsletter of The British Society for the History of Science, and the European edition of NASA's *Astrobiology* magazine, January 2007

A version of Chapter 2 appeared as 'Aliens and time in the machine age', by Mark Brake and Neil Hook, in *International Journal of Astrobiology*, **5**(4), October 2006, pp. 277–86.

A taster entitled 'Darwin's Bulldog and the time machine' appeared in the European edition of NASA's *Astrobiology* magazine, January 2007.

A taster entitled 'French tales of infinity' appeared in the European edition of NASA's *Astrobiology* magazine, May 2007.

A taster entitled 'The dreadful hammers of Jules Verne' appeared in the European edition of NASA's *Astrobiology* magazine, May 2007.

A taster of Chapters 3 and 4 appeared as 'Rocketry, film and fiction: the road to Sputnik', by Mark Brake and Neil Hook, in *Physics Education*, **42**, 2007, pp. 345–50.

A version of Chapter 7 appeared as 'Darwin to double helix: astrobiology in fiction', by Mark Brake and Neil Hook, in *International Journal of Astrobiology* (in press).

Readers should note that Chapters 1, 2 and 4 were written by Mark Brake; Chapters 3, 5 and 6 by Neil Hook. Chapter 7 was jointly written.

Chapter 1
A PLURALITY OF HABITABLE WORLDS: THE AGE OF DISCOVERY

On 15 March 1610, the rather improbably named Wackher von Wackenfels drove up in his coach to the house of the internationally famous scholar Johannes Kepler, in a state of great agitation. Wackher, amateur philosopher and poet, and Privy Counsellor to his Holy Imperial Majesty, excitedly told the mathematician of news that had just arrived at Court: a philosopher named Galileo in Padua had turned a Dutch spy-glass at the heavens, and discovered four new worlds.

A few days later, Kepler received further astounding evidence in the shape of Galileo's brief but magnificent flyer, *The Starry Messenger*. The book signalled the offensive on the feudal order of the old Universe with a new weapon of discovery: the telescope.

But Kepler had his own startling story to tell. Just one year before Galileo's earth-shattering discovery, he had published the first ever work of science fiction. Kepler's tale was a space voyage of discovery in the new physics that invented the alien and anticipated the Universe soon to be unveiled by the telescope.

In fiction and in fact, the Age of Discovery had dawned. By the end of the century, the layered Universe of Aristotle crumbled before the world machine of Newton...

A *new age*
This journey through the fantastic begins with the Renaissance. The mediæval rebirth in science and culture was one in which

Polish astronomer Nicholas Copernicus created science fiction, Johannes Kepler dreamt up extraterrestrials, and Irish satirist Jonathan Swift divined the dark side of the New Philosophy.

Science fiction is the narrative of change. It deals with the currents of change in science and society. There was no greater period of change than the age that saw the feudal world come crashing down. The causes were the new social forces and the Scientific Revolution. A new culture developed, capitalist in its economy, classical in its art, and scientific in its approach to Nature.

The voyages of discovery of the age led to space voyages in fiction. These early off-world narratives were explorations of alien life and alien worlds. They were written by a conscious vanguard of scholars and artists true to their time. They started with journeys to the Moon in Kepler's *Somnium* (*The Dream*) (1634) and *The Man in the Moone* (1638) by Francis Godwin, the Bishop of Llandaff in Wales. They morphed into the planetary novel with further flights of the fantastic, such as in Cyrano de Bergerac's *L'autre Monde* (*Other Worlds*) (1657).

All were influenced by a revolution in the ideas of science. This revolution was far greater than those in politics and religion, in that it contained the seeds for the indefinite advance of society. What the cost may be of this advance was the question posed by Jonathan Swift's *Gulliver's Travels* (1726), with its flying island of Laputa.

To trace the tale of this obsession with space voyages, and to look at these key examples of early science fiction, it is useful to consider how astronomy became a tool of empire due to the rapid powering of the Western economy by overseas trade. The greed for profit led to a speedy expansion of those unsung catalysts in the birth of modern science fiction and science – shipbuilding and navigation.

Discovery, piracy and plunder

The great sea voyages that started with the Portuguese explorers around 1415 opened the planet to European capitalist enterprise.

The voyages were the fruit of the first deliberate use of astronomy and geography in the pay of glory and profit. With a practical eye for sugar plantations, slaves and gold, a drive was begun for astronomical tables precise enough to be of use to the ocean navigators.

Theory and practice met at the Court of Prince Henry the Navigator at Sagres. Here, Moorish, German, and Italian experts planned voyages with hardened Portuguese and Spanish sea captains. The Turkish stranglehold on eastern trade raised the compelling idea of venturing into the Indian Ocean by some way other than the Red Sea. Strategists debated two promising alternative routes. The first, to round Africa, was profitably realized by the Portuguese in 1488, though India was not accomplished by Vasco da Gama until 1497.

The other route, volunteered by astronomers and geographers such as the Florentine Paolo dal Pozzo Toscanelli, was to head west over the uncharted ocean in search of China at the other side of the world. Now to reason such a hypothesis is one thing, but to sail straight out to sea quite another. In the popular imagination a host of potential fates might befall such adventurers. They might sail on without end, or they might plunge off the edge of the world. Not a single soul anticipated there would be an entire continent in the way.

The one man fit for such a gamble was a supreme navigator and adventurer. Christopher Columbus was many light years from being a scientist[1], but he had a brilliant vision. By seafaring west, he might find new worlds, or even discover 'a new heaven and a new earth'. It was this dream, part spiritual, part scientific, which finally enabled Columbus to seduce the venture capitalists.

There was a striking difference between the successive voyages of the Portuguese round Africa and of Columbus, wagering all to sail directly across the Atlantic. The successive voyages relied upon a gradual advance on the traditional route. The scientific proposal provided a revolutionary break with such tradition. True, Columbus may have had mystical motives. But the finance he received for his voyages was an investment in scientific potential: the promise of booty plundered from the proof of a theory.

It was this very spirit, personified by Columbus and blended with a heady brew of Copernicanism, which influenced the early works of science fiction. The unearthing of the New World and the ancient worlds of Asia, with their exotic goods and customs, made the classical world seem provincial[2]. Early science fiction authors and adventurers alike were won over by accomplishments that the classical world had never imagined. These architects of the Renaissance envisioned a new age. By the middle of the sixteenth century they had attained it. The discipline of navigation was rejuvenated, and more accurate maps were needed.

The movement of the heavens now had a cash value.

The world turned upside down

So it's no wonder that it was astronomy, so closely connected to geography, which brought the whole ancient system of thought crashing down.

For two millennia Aristotle's cosmology held sway. The Aristotelian cosmos was a two-tier, geocentric Universe. The Earth, mutable and corruptible, was placed at the centre of a nested system of crystalline celestial spheres, from the sub-lunary to the sphere of the fixed stars. The sub-lunary sphere, essentially from the Earth to the Moon, was alone in being subject to the horrors of change, death and decay. Beyond the Moon, in the supra-lunary or celestial sphere, all was immutable and perfect. Crucially, the Earth was not just a physical centre. It was also the centre of motion, and everything in the cosmos moved with respect to this single centre. Aristotle declared that if there were more than one world, more than just a single centre, the elements of earth and fire would have more than one natural place toward which to move – in his view a rational and natural contradiction. Aristotle concluded that the Earth was unique. There was no room for the alien.

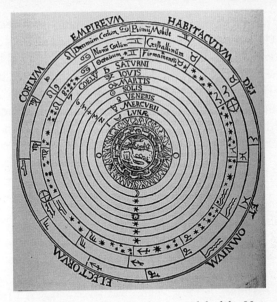

Figure 1.1 Aristotle's two-tier, geocentric model of the Universe, from the corrupt central Earth to the outer sphere of the Prime Mover (Library of Congress).

The Latin West, where Aristotle's physics was embraced and refined, took up his system in Christian teachings. Where Aristotle led, the Church followed. For Christians, the idea of alien life, or pluralism, directly disputed the notion of an omnipotent God. If God had wished to create another Earth, how could he do so without violating Aristotle's physics? The famous *Divine Comedy* (1308–21), by Italian poet Dante Alighieri, made it appear that the mediæval Universe could have had no other structure than Aristotle's, underlining the importance of the position of the Earth to the Christian drama of life and death in the Middle Ages. The vastest of all themes, the theme of human sin and salvation, is adjusted to the great plan of the Universe. Such would be the significance of Copernicus shifting the Earth from its 'natural' place at the 'corrupt' centre of the Universe.

So AD 1543 marks one of the great turning points in human history[3,4,5]. Copernicus' great book, *De Revolutionibus Orbium Coelestium* (*On the Revolutions of the Heavenly Spheres*)[6], placed the Sun at the centre of our planetary system, heretically downgrading the position of the Earth to that of mere planet. Copernicus set in train a revolution. A new physics was born, and a new mantra: if the Earth is a planet, then the planets may be Earths; if the Earth is not central, then neither is humanity[7].

Fly me to the Moon

Copernicanism caught fire.

For his belief in aliens, along with other alleged heresies, the Roman Catholic Church burned Italian mystic Giordano Bruno at the stake in 1600. Bruno had published a form of cosmic pantheism in his *De L'infinito Universo E Mondi* (*On the Infinite Universe and Worlds*) (1584). Bruno was passionately pluralist. He populated planets and stars, attributing souls to them, and even did the same for the Universe as a whole. In the words of Bruno: 'Innumerable suns exist; innumerable earths revolve around these suns in a manner similar to the way the seven planets revolve around our sun. Living beings inhabit these worlds'[8]. However, it was the very first work of science fiction, Kepler's *Somnium*, with which the spirit of Copernicanism first took flight.

Johannes Kepler was conceived at 04:37 on 16 May 1571, and born prematurely on 27 December at 14:30, after a pregnancy lasting 224 days, 9 hours and 53 minutes[9]. Such unerring accuracy, culled as it is from his own astrological charts, paints an immediate portrait of Kepler as a man of great contrasts and contradictions, typical of an age of transition. Kepler was said to stem from a noble family, but by the time of his birth the line had fallen into degenerates and psychopaths. Kepler's mother was raised by an aunt who was later burnt alive as a witch. His mercenary father narrowly escaped the gallows. An altogether curious pedigree, but perhaps telling for a man who was to become the most brilliant and erratic speculator of the Scientific Revolution.

In 1593 Kepler got his first job, as a Professor of Mathematics in Graz, Austria. It brought mixed rewards. Kepler was a brilliant scholar, but a rather poor teacher. Whenever he got excited, and he was almost always in this state, he 'burst into speech without time to weigh whether he was saying the right thing'[10]. His digressions constantly led him to think of 'new words and new subjects, new ways of expressing or proving his point, or even altering the plan of his lecture'[11]. Hardly any wonder that in his first year of teaching he had only a small handful of students, and in his second, none whatsoever. He was the very picture of an absent-minded unintelligible professor, delivering garbled lectures before an empty classroom.

Whereas Galileo was wholly and frighteningly modern, Kepler never severed himself from the mystical Middle Ages. Unlike Galileo, who was devoid of any spiritual leanings, Kepler was struck by the magical implications of the Sun-centred Universe. Kepler was fanatical about discovery. The spirit of scientific enquiry seemed to well up within him. His books on astronomy attempted to lay bare the ultimate secrets of the cosmos. Yet they were a hotchpotch of geometry, music, astrology, astronomy and the occult. Indeed, Kepler's famous three laws of planetary motion are buried deep within such a work of lavish fantasy.

This intoxicating mix of fact and the fantastic is also characteristic of his fiction. Kepler, in his darkest hours, was able to find refuge in his imagination – refuge from the death of three of his children and his first wife, and sanctuary when his detractors used *Somnium* against him when it was produced as key evidence in the trial of his mother for witchcraft. Writing, for Kepler, was a kind of cognitive labour, driven on by his thirst for knowledge. He was best known for his laws of planetary motion, and they alone make him a key figure in the Scientific Revolution. He was also a man of letters, many of which championed the cause of Copernicanism. Though his son Ludwig had *Somnium* published posthumously in 1634, Kepler had actually been working on the text since 1593, a mere 50 years after Copernicus.

From the very outset, science fiction had a subversive edge. Kepler selected the framework of *Somnium* to pass off his Copernican essay as a dream. In this way he was able to subvert the scorn of the Aristotelians by concealing his radical work in the guise of classical mythology. All of this was published initially in draft manuscript form in 1609, a year before Wackher von Wackenfels brought details of Galileo's visual proof of Kepler's flights of fancy.

Somnium is a truly extraordinary work. Its theme is the new Universe of Copernicus. There is seldom a passage in Kepler's twenty solid volumes of writings that isn't alive and kicking. So it is with *Somnium*. As a fictional travelogue it follows on from ancient Greek stories, such as Plutarch's *The Face on the Moon* and Lucian's facetiously titled satirical work *A True Story*. In all other ways it signals a sharp break with classical tradition.

The story of *Somnium* is an extrapolated voyage of discovery. Copernicus had shifted the centre of the Universe to the Sun. Kepler's aim was to explore this alien panorama from the alternative standpoint of the Moon. He wanted to describe what the new astronomy would be like from the perspective of another planet. In this way the feasibility of a non-geocentric system could be explored. Novel observations of the heavens, and the Earth itself, could be made. On his journey, the 'spirit of the Moon' transports Kepler's hero Duracotus to a lunar landscape. Once there, he gazes down upon the Earth looking at its geography and its motion through space, also exploring the surface of the Moon and its alien inhabitants.

Somnium's mastery of Copernicus is complete. Although the traveller in the book initially journeys to the Moon by spirits, the laws of physics then govern him, and science takes over from fantasy:

> the journey... becomes easier because... the body... escapes the magnetic force of the earth and enters that of the moon, so that the latter gets the upper hand.... At this point we set the travellers free and leave them to their own devices...[12]

In the *Astronomia Nova* (*A New Astronomy*) (1609) Kepler had come very close to the concept of gravity. In *Somnium*, he takes it for granted. With remarkable insight he also proposes the existence of zero gravity zones, '... for, as magnetic forces of the earth and moon both attract the body and hold it suspended, the effect is as if neither of them were attracting it...'. He takes a further step in the same direction by assuming that there are spring tides on the Moon, due to the joint attraction of Sun and Earth. The journey itself recognizes that the shortest route to the Moon is not the straight line believed by Lucian and Plutarch, but a trajectory from Earth to a point in space where the Moon and the lunar voyagers would arrive simultaneously[13].

Shortly, the speed of flight carries the travellers 'along almost entirely by our will alone, so that finally the bodily mass proceeds toward its destination of its own accord'. Kepler had developed the concept of 'inertia' and widened its remit to outer space. He also keenly understood the dangers of lunar flight. He believed it was scientifically possible for men to reach the Moon, again distinguishing his work from the fantasy utopian writers of the past[14].

Lunar nightmare

The journey done, *Somnium* then unveils Kepler's imagined Moon. It is a picture painted with characteristic Copernican care. From sunrise to sunset, a Moon day lasts about two weeks. So too does a lunar night, as the Moon spins on its axis once a month. Since it also takes a month to carry out an orbit, the Moon always shows the same face to the Earth. The Moon creatures know the Earth itself as 'volva' (as in *revolvere*, to turn). The Earth-facing half they call the Subvolvan, the far side is the Privolvan; both halves have a year of twelve days and nights.

Kepler now reveals the nightmare of a truly alien world. Deadly differences in temperature – blazing days, freezing nights – plague the landscape. The starry heavens above are equally strange and unfamiliar. Across a bible-black sky, the stars, Sun and planets scurry relentlessly back and forth, due to the Moon's trajectory around the Earth. This captivating 'lunatic' astronomy of Kepler's is yet to meet its match in fiction.

On the Moon itself, the picture becomes more desolate. The extraterrestrials that stalk Kepler's lunar world are not men and women, but creatures fit to survive the alien haunt. Two centuries before Darwin, Kepler had grasped the bond between life forms and habitat[15]. During his time at university, Kepler had written:

> If there are living creatures on the Moon (a matter which I took pleasure in speculating after the manner of Pythagoras and Plutarch in a disputation written in 1593), it is to be assumed that they should be adapted to the character of their particular country[16]

The Privolvans have it the hardest. There is no let up in their long nights 'bristling with ice and snow under the raging, icy winds', and no respite provided by Earthlight, which they never see. Their 'day' is no better: for two weeks a scorching Sun bakes the lunar air so that it is 'fifteen times hotter than our Africa'. The Privolvans have 'no fixed and safe habitations; they traverse in hordes, in a single day, the whole of their world, following the receding waters either on legs... or on wings, or in ships'. In the Subvolvan hemisphere, the nights are softened by the light and heat of the Earth, which hangs motionless in the sky 'as if nailed on'. The Subvolvans enjoy the great spectacle of the waxing and waning of the Earth's surface, fifteen times the size of our Moon.

The mountains of the Moon are higher than those of the Earth. All of Kepler's creatures attain a monstrous size. They feed only at night, for feeding after sunrise often leads to death. The hide of these massive serpents is permeable and, if exposed to the full force of the Sun, becomes seared and brittle. Such is life for this gigantic race of short-lived beasts. They bask for a fleeting instant in the rising or setting Sun, then creep into the impenetrable lunar darkness[17].

Somnium is a watershed. It marks the end of the old era, and heralds the arrival of the new science. Its hypothesis on extraterrestrial life had an estranging effect, revealing the world in a new light. Kepler's book is an imaginative *tour de force*. It was the first space fiction of the age. The alien voyages quickly evolved, from Francis Godwin to Jonathan Swift, as a potent motif for exploring the insignificance of man.

Somnium's influence was huge, inspiring other interplanetary journeys such as those of John Wilkins, Henry More, H. G. Wells and Arthur C. Clarke. Kepler was a pioneer of the new vision of deep space as the home of a plurality of inhabited worlds[18]. There is no greater testament to the power and imaginative sway of science fiction than this. Despite the incredible odds against detecting life beyond our planet, billions have been spent in the twentieth century on sober scientific projects in the search for alien intelligence. That search started with Kepler.

The planets may be earths...

It was the telescope that eventually shattered Aristotle's crystalline Universe. Italian mathematician Galileo Galilei wielded the newly invented device like a weapon of discovery. A new Universe was unveiled. Kepler's fictional vision was finally writ large in the sky. Earth-like mountains and craters on an imperfect Moon. Impure spots on the Sun. Four moons in orbit around Jupiter, a focus of gravitation other than Earth. Countless stars that could only be seen with the aid of a spyglass. So much for the perfection and immutability of Aristotle's heavens.

Not everyone believed Galileo's discoveries with the 'optick tube'. Copernicus had been a quiet, unassuming man all his life. No one who met the captivating Kepler could seriously dislike him. Galileo was another matter. Portraits show a ginger-haired, thickset man with rough features and an arrogant stare. Galileo held radical leanings and a contempt for authority that would later lead to infamous trouble. His exceptional talent for antagonizing others led to scepticism in the small academic world of his own country.

Figure 1.2 Drawing of the Moon from Galileo's *Sidereus Nuncius*, 1610.
Galileo urged the reader to imagine walking over mountains and craters
to convey its Earth-like nature, contrary to Aristotle's teachings

On the evenings of 24 and 25 April 1610, a memorable party
was held to celebrate Galileo's recent discoveries. The great man
was invited to demonstrate the Jupiter moons in the spyglass. Not
one among the eminent guests was convinced of their existence.
The crude nature of the mysterious gadget didn't help, but many
were blinded by prejudice. They refused to look down the tube.

Galileo's arrogance aroused uproar. A public controversy followed similar to the UFO debacle three hundred years later.
There were claims of optical illusions, haloes, and the unreliabil-
ity of inexpert witness. As Galileo's discoveries had become the
talk of the world's poets and philosophers, scholars in his own
land were sceptical or downright hostile. Only Kepler's weighty
voice was raised in defence.

The Scientific Revolution begins with the discovery of these
other worlds, symbolized by the names of Kepler and Galileo.
They produced a map of the knowable, just as the unknown was
at the point of becoming known. Galileo's revolutionary pam-
phlet *Sidereus Nuncius* (*The Starry Messenger*), written in 1610 in

Italian rather than pompous, scholarly Latin, describes shadows cast by lunar mountains and craters.

Galileo implied that Kepler's creatures might well be dwelling there. It was a vital new piece of evidence in the debate on the existence of extraterrestrial life. It is the first time the Moon becomes a real object for us. At that same instant we feel a sense of wonder, or *estrangement*, from this new reality. Estrangement implies a state of imperfect knowledge. It is the result of coming to understand what is just within our mental horizons[19].

The birth of science fiction

This same sense of wonder is common to science fiction. Little surprise then that Galileo inspired Kepler to argue that if Jupiter has satellites (a word coined by Kepler himself) then it also must be inhabited. Indeed, it was Kepler who motivated H. G. Wells to write, nearly 300 years later, 'But who shall dwell in these Worlds if they be inhabited? Are We, or They, Lords of the World? And how are all things made for Man?'[20]

Galileo had to assume that shadows on the Moon have similar causes to shadows on Earth in order to understand the Moon's difference from Earth. Yet so great an astronomer as Kepler evidently needed to believe in extraterrestrials in order to render Galileo's discovery thinkable. Kepler realized that to understand the Moon it was not enough to put one's observations into words. The words themselves had to be transformed by a new sort of fiction. That's why there is something revolutionary about *Somnium* in the history of science. Throwing words at the Moon, as it were, has a dialectic effect – the words come back to us changed. By imagining strange worlds, we come to see our own conditions of life in a new perspective.

Kepler's journey to the Moon is the first work of science fiction. Others have argued for Lucian's *A True Story*, written a millennium and a half earlier. The striking difference was this. Lucian's work is fantasy. Kepler's *Somnium* was a conscious effort to understand the new physics. The key factor in developing this consciousness was the evidence provided by the newly invented telescope. Progress in science then drove continuity in its fiction.

The change in the economy had other effects on the material conditions of Kepler's time compared to Lucian's. The development of the printing press had transformed the spread of knowledge. There was an increasingly diverse readership, and publication rates rocketed. In England, for instance, only eighty books were published per year in the 1540s, but by the 1640s this figure had exploded to four thousand. No doubt Kepler was aware of the growing use of print as an ideological battleground. His dream tale suited the traditional oral culture. By the 1620s the telling of sensationalist stories with a moral purpose had become a very popular form.

Science fiction began with the Scientific Revolution. It marks the paradigm shift of the old Universe into the new. Aristotle's cosy geocentric cosmos was about us. The new Universe of Kepler and Galileo was decentralized, inhuman, infinite and alien. Historically then, science fiction is a response to the cultural shock created by the discovery of humanity's marginal position in a Universe fundamentally inhospitable to man[21].

Science fiction is our attempt to make human sense of Copernicus' new Universe.

Copernicus, the bishop and the wild goose chase

As Kepler was expanding the Copernican dominion, the Bishop of Llandaff in Wales, Francis Godwin, was writing the first space travelogue in the English language. Godwin's action packed *The Man in the Moone: or, A Discourse of a Voyage Thither, by Domingo Gonsales*, is another crucial work of early science fiction.

Published posthumously in 1638, the Bishop started to write the book in 1589. He didn't live to see its outstanding success. *The Man in the Moone* went through two dozen editions, well into the eighteenth century, and was translated into many languages, including French, Dutch and German. Godwin's novel was seen as the archetypal space voyage for the next hundred years or so.

Even nineteenth century writers such as Jules Verne and Edgar Allan Poe credit him as a significant inspiration.

Godwin's journey captured the imagination of John Wilkins, First Secretary of the Royal Society. Wilkins had published the factual *The Discovery of a World in the Moone* (1638), which proved a fashionable piece of science communication. The book was Wilkins' contribution to the debate on the plurality of habitable worlds. The third edition was revised to take account of the popularity of Godwin's work. The newly added fourteenth chapter ends:

> Having thus finished this discourse, I chanced upon a late fancy to this purpose under the fained name of Domingo Gonsales, written by a late reverend and learned Bishop; In which (besides sundry particulars wherein this later chapter did unwittingly agree with it) there is delivered a very pleasant and well contrived fancy concerning a voyage to this other world[22]

Wilkins, and Godwin himself, considered the main point of Godwin's book to be an account of the possibility of a space voyage to another world. Equally important, however, is this: *The Man in the Moone* is the first English book in history to portray alien contact. Like Wilkins, Godwin believed contact to be a real possibility; it was only a matter of time before such an encounter took place. Details of Galileo's lunar discoveries with the telescope are also clearly evident in the book. The preface of Godwin's fiction credits Galileo as the discoverer of this new world.

Godwin had been born in 1562, the son of Thomas Godwin, Bishop of Bath and Wells. It was during Godwin's time at Oxford that Giordano Bruno delivered his sensational debates on the question of life beyond the Earth. It may be that Bruno inspired the idea in Godwin of an inhabited Moon, though it was through Galileo that the notion became widely known.

Whilst at Llandaff, Godwin had lucrative connections with the thriving local sea trade. Captains and merchants out of the ports of Cardiff and Bristol dined with Godwin at the Bishop's Palace in Llandaff, a mere three miles from the Cardiff docks. Even though Godwin was entertaining for funds to restore the Cathedral, he also heard maritime tales of discovery and endeavour. His fiction is similar to the diaries kept by notable explorers of his time, such as Richard Hakluyt, a style that he may have been trying to copy.

Godwin's *The Man in the Moone* features a protagonist, Domingo Gonsales, whose voyage of discovery goes astray to St Helena, the Moon and eventually China. During his travels Gonsales rears and trains forty wild geese ('*gansas*') as a bizarre flying machine. On his return to Spain his ship is attacked and sunk by British privateers. Attempting to escape, Gonsales harnesses himself to the birds, which carry him higher and higher. Soon enough he remembers they migrate each year to hibernate on the Moon. His interplanetary flight is full of astonishment. The greatest surprises confirm the Copernican cause and the new

Figure 1.3 Science influenced by science fiction: John Wilkins' *The Discovery of a World in The Moone*, published in 1638.

physics. The Earth is not the single centre of gravity in the Universe, and rotates on its own axis:

> when we rested... either we were insensibly carried (for I perceived no such motion) round about the Globe of the Earth, or else that (according to the late opinion of Copernicus) the Earth is carried about, and turneth round perpetually, from West to East, leaving unto the Planets onely that motion which Astronomers call naturall[23]

On leaving the Earth, Gonsales gets progressively lighter, and then heavier on reaching the Moon. It's a clear example that Godwin too was trying to convey the principle of gravity. It seems that, in fiction and in fact, the idea of gravity had come of age. Flirted with by Godwin and Kepler, it was finally established in Newton's *Principia* (1687).

After a flight of twelve days, Gonsales presently sets down on a hill. He finds the Moon a world like our own, but on a much bigger scale. People, plants and animals all reach gigantic proportions. Indeed, in the hierarchy of lunar society stature is a sign of nobility. The 'true lunars' are thirty times taller than humans, and live up to thirty times longer. They lead an idyllic life. In contrast, the 'dwarf lunars' are little taller than humans, live no longer than eighty years, and are consigned to the most menial tasks. Given this status quo, Gonsales is resigned to considering himself rather inferior.

Godwin describes his lunar world as a utopia, but the details make you doubt it. True, even the 'dwarf lunars' are more virtuous than their terrestrial cousins. The most mortal wound can be cured. Crime is unknown. Lunar women are so beautiful that no man wants to commit adultery. Soon, however, it's disclosed that the 'true lunars' practice a radical form of eugenics. Defects are identified at birth. So, since the lunars don't kill, the newborn potential sinners are shipped Earthwards, to North America.

Godwin's *The Man in the Moone* is another key work of early science fiction. The idea of alien life was new and exciting, and thanks to Kepler and Galileo the possibility of extraterrestrials seemed very real for the first time. So much so that, with the benefit of over-eager hindsight, the idea was mistakenly credited to Copernicus.

Godwin dared suggest that extraterrestrials might not only be superior to us; they might also be happier. Kepler's *Somnium* had made alien existence seem like a nightmare compared with life on Earth. Inventing the alien contact story, Godwin hinted at an early form of evolution. The Universe may harbor a more highly evolved race than man. It was an idea that would only become commonplace two hundred years later.

Empires of the Sun and Moone... Cyrano de Bergerac

The space travel of Kepler's 'Moon spirits' and Godwin's wild goose chase soon developed into the invention of the rocket. Furthermore, in one generation the space voyage developed into a more critical 'contact' fiction, typified by that of Cyrano de Bergerac.

Cyrano, notorious French duellist, satirist and freethinker, had a life immortalized by many romantic legends. Many of these legends derive from 1898, when Edmond Rostand cast Cyrano as the title hero of a play. The real life counterpart easily equalled the legend of the swashbuckling swordsman with a large snout. Cyrano allegedly fought with the elite Gascon Guard, being discharged through serious injury. One of the circle of dramatist Molière, Cyrano was a radical who liked to parade his unorthodoxy in outrageous style.

Cyrano had studied with philosopher and mathematician Pierre Gassendi. Gassendi's life and work was committed to changing the perception of science. He was admired as a successful science communicator, which in no small part was due to the influence of his pupil. Cyrano's reputation rested on the *L'autre Monde* (*Other Worlds*) trilogy: *Les États et Empires de la Lune* (*The States and Empires of the Moon*) (1657), *Les États et Empires du*

Soleil (*The States and Empires of the Sun*) (1662) and the lost work, *The History of the Stars*.

Copernicus, Galileo and Kepler had been pious men. Not Cyrano. Bishops Godwin and Wilkins had believed in life beyond the Earth as further proof of God's majesty. Cyrano's view on alien life was uncompromisingly atheist. Under the influence of Descartes, he banished God from the heavens, using only reason as his guide.

Cyrano caught the mood of his age. His books were astonishingly successful. They were popular throughout Europe, and a great influence on satirists such as Voltaire and Swift. So strident were the first two volumes of his trilogy that the early editions had their heretical contents toned down.

Cyrano took positive delight in scandalizing the Church. The atheist views implicit in the new science were openly developed in his science fiction. The notion of a Universe fit for life was stripped of any divine overtone. Cyrano didn't assume his worlds to be the work of a Creator. His space novels were an unprecedented attack on human self-esteem, designed to create utter disillusionment.

Once more it is striking to see how the militant implications of science were firstly realized in fiction. Descartes' physics suggested an un-Christian view of alien life, which Descartes was wary of confronting in public. Cyrano's daring tackled the topic head on. The first two parts of *L'autre Monde* spoke of imagined meetings between man and alien.

Cyrano conjures up a cosmos in which there are as many planets as there are stars. These countless worlds, like the analogous Moon and Sun in his story, are inhabited by rational beings. A Universe teeming with intelligence is only his starting point. Cyrano then uses this setting to look at the uncertainty of man's nature and status. His findings are not reassuring. The novels present the most potently critical view of the pre-eminence of humanity, and of religion, than any other work of their age.

A world of baboons and plucked parrots?

In *The States and Empires of the Moon*, the narrator is a man named Dyrcona, an anagram of d[e] Cyrano. Dyrcona initially travels by the ingenious use of bottled dew. Fastened around his waist, the evaporating dew lifts our hero to the Moon. His mission aborts. He falls to Earth, initially to the French colony of Canada, opened by Jacques Cartier only a century before. On capture, the Viceroy quizzes Dyrcona. Curiosity focuses on his means of travel, and the new astronomy. Dyrcona replies, 'I believe that the planets are worlds surrounding the sun and the fixed stars are also suns with planets surrounding them'[24]. The Viceroy concludes that the Universe is infinite and Aristotle wrong. By attaching fireworks to a travelling machine, our hero now rockets to the Moon.

According to Arthur C. Clarke[25], de Bergerac is to be credited for inventing the ramjet. Cyrano wrote:

> I foresaw very well, that the vacuity... would, to fill up the space, attract a great abundance of air, whereby my box would be carried up; and that proportionable as I mounted, the rushing wind that should force it through the hole, could not rise to the roof, but that furiously penetrating the machine, it must needs force it upon high[26]

Dyrcona finds the Moon to be a world in every sense of the word. Like Godwin's vision, it is the home of a race of giants. Cyrano's lunar men, the Lunarians, walk on all fours, and greet with disbelief this tiny bi-ped from their 'moon'. Even the lunar beasts walk on all fours, highlighting further Dyrcona's absurd appearance. So humans are not comparable with the animals of the Moon, let alone the extraterrestrials.

The Church's claim to man's sole custody of reason and immortality is made to look absurd. After a sparkling account of an atheist genesis of life and the Universe, Cyrano cites rationalism as the main feature of the enlightened lunar culture.

The highly intelligent Moon-dwellers feed on scents and slumber on flower blossoms. They use poetry as a form of tender, and are convinced the stars are inhabited. They see death as the fulfilment of life. Humanity's belief in God is viewed merely as a defect of reason.

The inferiority of humans is further exposed in a series of weird events. Dyrcona is trained to perform tricks for the Lunarians, just as apes may be trained on Earth. He then meets Gonsales, from Godwin's novel, and finds that Gonsales is thought to be a baboon. He is kept as a laughable and bizarre pet, and shown in a zoo. In fact, the conversations between Dyrcona and Gonsales while in the zoo are particularly amusing to the Lunarians. Once the gossip starts that the two caged bipeds may be flawed 'humans', the church steps in. An edict is decreed. Any comparison of the earthly creatures to 'men', or even lunar animals, is blasphemy.

Dyrcona finally stands trial. The charge: believing that his own planet is a world, and not a 'moon'. The Lunarians argue for the total insignificance of humans. Indeed, the bi-peds are thought to be monsters devoid of reason, and are best described as 'plucked parrots'. Ironically, the court's decision is still to declare Dyrcona a 'man' so he can be charged with the degrading punishment of recanting his beliefs at every street corner on the Moon.

Set the controls for the heart of the Sun

In *The States and Empires of the Sun*, man is further demoted. Dyrcona flies sunwards in an ingenious flying machine powered by solar reflectors. Once his four-month voyage is over he sets down on a sunspot. He discovers a strange solar world of wonderful colour. The Sun is a globe of 'burning snowflakes' and plant life of stunning beauty. Dyrcona watches as living creatures constantly morph into birds, trees and human beings, until they are finally revealed as beings more perfect than man.

Eventually, Dyrcona is once more put on trial. In the section *Histoire des Oiseaux* (*Story of the Birds*), the Sun is revealed in all its glory. It is the realm of the souls of all the creatures in the cosmos. It is the region of truth and of peace. It is the province of lovers. It is the state of philosophers. Mostly, however, it is the abode of the birds.

Dyrcona is captured and tried by the birds. Cyrano now develops the theme of man as a plucked bird. To the wise and cultured birds, man is seen as a deluded creature. He believes the entire animal world, and the environment, to be at his disposal. He is a featherless monster, an instigator of conflict and a brutal destroyer. In an attempt to trick his judges, Dyrcona pretends he is an ape. But the birds are not fooled. They decide that man is a bane of existence 'of which every well-policed state should be rid'.

Cyrano's Sun and Moon are worlds like our own. Cyrano makes his solar birds as arrogant as man. They too believe themselves the most supreme, most rational and most cultured beings in the Universe. They are equally cruel. They condemn Dyrcona to a gruesome death to be eaten alive by flies, a fate he escapes. The Lunarians treat the humans as circus monsters in the same way that humans would if the Moon dwellers came to Earth. The Lunarians enjoy dressing Domingo Gonsales as a pet monkey. Gonsales himself, with equal conceit, declares to the Moon dwellers that all life in the Universe is created for Spaniards to enslave.

Cyrano satirized the bigotry and beliefs of the age. His masterpiece, *L'autre Monde*, was the most secular picture of the potential meaning of a Universe fit for life. It was a Copernican stand against the anthropocentric notion that mankind was the centre of creation.

The combined efforts of Kepler, Godwin and de Bergerac ensured that every important cosmology since has held the idea of alien life as a basic principle. In 1686, for example, Bernard le Bovier de Fontenelle[27] planted aliens deeper in the European psyche with his *Conversations on the Plurality of Worlds*, in which he used Copernican perspectives in support of extraterrestrialism[28].

Before going on to consider the development of the planetary novel with Jonathan Swift, it is useful to look at the way in which science drove a change in language. The campaign for progress produced its first dark philosopher in Francis Bacon.

Science communication and the new philosophy

The science fiction of Kepler, Godwin and de Bergerac played a major part in the communication of the New Philosophy. Ultimately, the proof Galileo gleaned with the telescope could not be ignored. It led to his infamous trial. It gave enormous prestige to the new experimental science. Before the revolution in ideas could be felt in practice, however, its potential needed to impact on the progressive class of adventurers, politicians and capitalists.

Science communication was born. Copernicus had been so terrified of the religious response to *De Revolutionibus* that he had delayed publication until he lay on his deathbed. Even then the tome's preface had been stamped with the esoteric motto 'For Mathematicians Only'. Hardly a call to arms, and little surprising that the work came to be known as 'the book that nobody read'.

In great contrast we have Galileo and Kepler.

Galileo's short but momentous book *The Starry Messenger* hurled his telescopic discoveries like a bomb into the arena of the learned world. It was printed in Italian, and in a new, tersely written style, which no scholar had used before. The result was culture shock. The sophisticated Imperial Ambassador in Venice said the book was 'a dry discourse or an inflated boast, devoid of all philosophy'[29]. Nonetheless, the book created the dramatic effect for which it was designed. It aroused immediate and passionate controversy.

Although Kepler's laws of planetary motion figure in orthodox histories of science, it is *Somnium* that gives us the true essence of Kepler. New worlds open to description are sometimes called the *novum*. The *novum* cannot be understood without inventing stories through which we are better able to comprehend. Kepler's innovative fiction satisfies our desire to understand something more than the mere data of our experience. Science fiction, starting with Kepler, promises communicable, rational discovery, not the fruits of religious revelation. It is critical and reasoning, as well as creative.

A new tongue was needed for science. The style of the English language was drastically simplified in the seventeenth century[30,31], and the new philosophers were quick to adapt. The science fiction stories of Kepler, Godwin and de Bergerac were short narratives of a few dozen pages. The first manifestos of science came from the earliest scientific organization, the Royal Society. They vowed to put an end to 'the luxury and redundance of speech' and to:

> ... reject all amplifications, digressions, and swellings of style: to return back to the primitive purity, and shortness, when men deliver'd so many things, almost in an equal number of words... a close, naked natural way of speaking; positive expressions, clear senses; a native easiness: bringing all things as near the Mathematical plainness, preferring the language of Artizans, Countrymen, and Merchants, before that, of Wits, or Scholars.[32]

Bacon's New Atlantis

The key prophet and publicist of the New Philosophy was Francis Bacon. Bacon was a kind of Renaissance spin doctor. He stood at the cusp of mediæval and modern science. His radiant vision was to put scientific ideas into practice, which he was determined to showcase to the world.

Bacon spun the application of science. He believed in science's promise to trigger a more rational world. He was hooked on pinning down Nature – amassing materials, executing experiments on a massive scale, and drawing conclusions from the sheer weight of evidence. His was an essentially *inductive* method. Bacon held that, given an army of well-drilled research workers, the mass of facts would ultimately lead to the truth. His goal was to shape an *organization* of philosophers that could direct the forging of new systems.

Science fiction was used to further the cause. Bacon's project found its highest praise in his utopian work *New Atlantis*. A group of merchant adventurers on their way from Peru to China find themselves blown off course. They're washed up on the shores of a mysterious civilization known as Bensalem, an idyll in which the inhabitants care for each other with virtually no civil problems.

Key to the narrative is Bacon's fictional House of Solomon, a kind of generic laboratory[33]. The fiction was based on the laboratory of the Danish nobleman and astronomer Tycho Brahe at Uraniborg. The House of Solomon motivated the creation of the Royal Society.

The scientists of the House of Solomon make huge progress using the inductive method. They covertly collect all the knowledge of the world on a twelve-yearly cycle. Their spies stalk the globe collecting data, returning once more to Bensalem. Solomonite scientists take this knowledge and turn it into concrete scientific advance. The science fiction of *New Atlantis* advocates secular politics and the use of science as a tool of the state[34]. The book undermines theology. It rejects any idea of the 'heavenly city', and attempts to recast knowledge as the pursuit of the creation of a 'heaven upon earth'.

The dark side

There was trouble in utopia. The New Philosophy had a dark side, which satirists were swift to sense. The paradox of the Scientific Revolution was this. Those who contributed most – innovators from Copernicus to Bacon – were also the most conservative: conformist in religion and conventional in philosophy. If they were not orthodox it was only because they believed orthodoxy had wandered from the path of reason.

Bacon believed that organized science would forge material progress[35]. A progress cleansed of the insidious notions of the Ancients. But he fell foul of his own warnings. He developed a flawed mediævalist ideology of power with his idea of

'Monstrosity'. People, Bacon said, had 'degenerated from the laws of nature'[36]. They had taken 'in their body and frame of estate a monstrosity'[37]. He drew his hypothesis from classical antiquity, the Bible and recent history. Among the multitudes who deserved destruction were, 'West Indians, Canaanites, pirates, land rovers, assassins, Amazons and Anabaptists'[38].

Bacon foresaw a utopia, and empire, of science. He claimed to be seeking to 'enlarge the bounds of human empire to make all things possible'[39]. This will to power, however, violently crushed alternatives. The expropriating British Empire expanded. Native colonial peoples were dispossessed, shot, poisoned and diseased. All this was made more possible by the ideology of Bacon.

The New Philosophy claimed to divide the Universe into two parts: one physical, and one moral. The primary physical truths were open to science. The moral domain lay in the realm of revelation[40]. Science fiction was free to explore the passions, while science allegedly worked free of moral concerns. Fiction braved the political sphere, where science feigned a cautious tread.

Bacon seized the idea that understanding Nature was also the means of taming it for profit, but there is a great deal of difference between an idea and an achievement. The New Philosophy, in the guise of the Royal Society, promised far more than it could deliver. There was good reason for the hostile reception it got from its critics.

The flying island of Laputa

The most famous critic was Jonathan Swift, whose satire *Gulliver's Travels* (1726) was named by George Orwell as one of the six most indispensable books in world literature[41]. Swift's novel is a skit on science and human nature, and a parody of the contemporary traveller's tales. It is also the best illustration of the eighteenth century planetary novel, which often took the form of a satire.

The science fiction of *Gulliver's Travels* is distinctive, particularly the voyage to the flying island of Laputa. After pirates have taken over his ship in the South China Sea, Gulliver drifts ashore to a landmass called Balnibarbi. Hardly has he set foot on land when a 'flying island' appears.

Made mainly of metal, and measuring four and a half miles in diameter, the floating island is magnetic. Buried inside the island is a six-yard long bi-polar magnet. It is held in abeyance by an intense magnetic field below the Earth's surface at Balnibarbi. The Laputian magnet may be turned so that the inhabited island can be steered with precision, horizontally or vertically, but only within the scope of Earth's magnetic field, which Swift estimates at four miles.

Swift had derived the idea for this floating island from the magnetic experiments of English physician William Gilbert. It is interesting to speculate whether Swift's island is a planet or an advanced technology. The island has a layered geology like a planet. The landing steps and observation galleries, however, suggest an airship. The attraction across the gulf of space between Laputa and Balnibarbi implies a planetary model. The problem, though, is that Gilbert had argued the Earth to be a magnet. So the decisive force in Swift's limited two-planet model would be magnetism, not Newtonian gravity. Finally, the fact that Laputian scientists can engineer the position of the flying island greatly weakens its likeness to a planet.

So perhaps Swift's flying island is science fiction's first spaceship: a space colony of technologically superior 'people'. It's possible that Swift is poking fun at contemporary efforts to build airships. One such notional airship, proposed in the seventeenth century by German Jesuit scholar Athanasius Kircher, based its design on the same magnetic principle.

Even so, there is sufficient evidence to suggest that Swift is toying with the idea of contact with a higher intelligence. The 'race of mortals' Gulliver meets on Laputa are the most alien he encounters on his travels. With heads cocked, and with one eye looking inward as the other scans the zenith, the intellect of these 'men' is constantly occupied. Draped in garments adorned with suns, moons and stars, they absent-mindedly speculate on mathematics and astronomy. These semi-crazed researchers, the first appearance of the mad scientist in fiction, have to be roused from their reverie by means of a rattle. Understandably their wives have run off.

Every facet of life on Laputa is regulated by science and mathematics. The beauty of an object is judged by its geometry. Meat is dished up in geometric shapes. Clothes are tailored by compass and quadrant. Swift is clearly mocking the science of his day. Its redundancy and irrelevance is ridiculed; its failure to create a practical payback lampooned.

But the scientists of Laputa are nevertheless superior to their English counterparts. Gulliver represents a world, Newton's world, remedial in science, particularly astronomy and mathematics. The Laputians show little curiosity in the affairs of Earth, save a single terrestrial member of their court who is viewed as 'the most ignorant and stupid person among them'.

With his portrayal of the Laputians, Swift furthers the idea of alien contact. As with Godwin and de Bergerac, a powerful extraterrestrial civilization meets with lowly Earthlings. Swift, however, was swimming against the stream. The majority of eighteenth century planetary novels were voyages of human conquest in space. Swift has the aliens come to Earth. The lofty and tyrannical Laputians are not the benevolent guardians of Steven Spielberg's *Close Encounters of the Third Kind* (1977). They are the brutal aliens of H. G. Wells' *The War of the Worlds* (1898) and countless pulp fictions of the future.

Laputa dominates the country above which it soars. Like some of the more disreputable spaceships of twentieth century fiction, any protest is punished. By manoeuvring Laputa the land below can be deprived of sun and rain. A further reckoning may be exacted. Any discontents can have their towns subject to missile attack, or destroyed completely, by having Laputa itself plummet to Earth. The ominous presence of this superior civilization above the Earth is a satire on the inhumanity of science. It is also an early example of the fear of an alien menace.

Swift's narrative of flying islands and mad scientists was an enormously influential book, and sublime in the effortless way in which it conjured up alien cultures. He vented his wrath at the claims, such as those by Bacon, of the right to subjugate others. He also attacked the use of violence in the name of progress:

> A crew of pyrates... go on shore to rob and plunder; they see a harmless people, are entertained with kindness, they give the country a new name, they take formal possession of it for their king... ships are sent with the first opportunity, the natives driven out or destroyed, their princes tortured to discover their gold...[42]

Swift is damning in his verdict, 'this execrable crew of butchers employed in so pious an expedition, is a modern colony sent to convert and civilise an idolatrous and barbarous people'.

Figure 1.4 Gulliver and the floating island where the Laputians run their world through mathematics and science.

He mercilessly mocks, 'the strange experiments of the scientists of the Royal Society'[43]. The tiny Lilliputians signify those whose self-importance reduces them in size. The Brobdingnagians are practical but cannot comprehend abstractions. The flying Laputians soar above the realities of life in their flying island, and the Balnibarbians are pedants obsessed with their own field, ignorant of everything else.

At times Gulliver discovers seemingly ideal communities modelled on the utopian concepts of fifteenth century English statesman Thomas More. Ultimately, however, Swift remains sceptical. Such utopias favour the collective over the individual. Whenever mankind designs an ideal community it takes for granted the idea that there is no limit to human understanding. Yet the scholars of Balnibarbi, who attempt to extract sunlight from cucumbers, show there should always be limits to mankind's explorations in science.

Swift sought to reassert the moral context of science. He challenged the New Philosophy with its physical and moral division. Swift's comment 'There are none so blind as those who will not see'[44] in his massively impolite *Polite Conversations*, summed up his disdain for scientists. Their blind pursuit of knowledge unfettered by caution was in danger of plunging the world into chaos.

The system of the world

As the Age of Discovery drew to a close, the one great practical success of the new science lay in navigation. The assumed superiority of European civilizations over the old worlds of Islam and the East was down to scientific technique. The new philosophers felt superior to their associates of the uncivilized Middle Ages, and even to the legendary ancient Greeks. Moderns, it was thought, might not be wiser, but they were undoubtedly more ingenious. They could achieve things the Ancients never imagined, like shooting off guns or sailing to the New World.

It was only a beginning. Already in 1619 utopian writer Johan Valentin Andrae had declared, 'It is inglorious to despair of Progress'[45]. That vision, so alien to the mediæval, if not totally to the classical, mind, let fly, subsuming everything in its path[46,47]. Yet science fiction still stood in its way.

The Copernican Revolution, and its inspired space voyages of Kepler, Godwin and de Bergerac, bore witness to the supreme triumph of the Age of Discovery: a cosmology that accounted for the motion of the heavens. In science, as in politics, the shattering of traditions meant that human creativity began to illuminate those disciplines where previously there was darkness. No part of the Universe was too distant for the interests of the new science[48], no narrative too fantastic for its fiction.

The drive for profit made astronomy a tool of empire. The invention of the telescope drove the imagination, forming a fiction that was creative, critical and reasoning. The new space voyages were increasingly used as a political tool. Kepler advanced Copernicus in the face of the Inquisition. Cyrano militated atheism at a time when the State reigned supreme. Swift despaired of the dark and dubious aspects of progress.

The new philosophy's ultimate expression was Newton's *Principia*, and its theory of universal gravitation. *Principia*'s triumph represented the collective effort of many great workers of the age – Copernicus, Galileo, Kepler, Descartes, Hooke, Huygens, Halley and Wren. Newton's work was to influence generations more over the next two centuries. In sustained development of physical argument, *Principia* is unrivalled in the history of science. In its physical insight and effect on ideas, only Darwin's *Origin of Species* compares. Newton's grand object was to demonstrate how universal gravity could maintain the system of the world. The Moderns were effectively consigning the ancient system to the dustbin of history. Science fiction would have to face the daunting prospect of a world in which man was the measure of all things. The dawn of the Mechanical Age had come.

References

1. Hakluyt Society (1930) *Select Documents Illustrating the Four Voyages of Columbus* (ed. C. Jane), series II, vol. 65, London.

2. Taylor, E. G. R. (1934) *Late Tudor and Early Stuart Geography 1583–1650*. Methuen, London.
3. Koyré, A. (1961) *La Révolution Astronomique*. Paris.
4. Kuhn, T. S. (1957) *The Copernican Revolution*. Harvard University Press, Cambridge, MA.
5. Koestler, A. (1959) *The Sleepwalkers*. Hutchinson, London.
6. Royal Astronomical Society (1947) *Nicolaus Copernicus, De Revolutionibus*, Preface and Book I (transl. J. P. Dobson). *Occasional Notes*, No. 10, London.
7. Dick, S. J. (1998) *Life on Other Worlds*. Cambridge University Press, Cambridge.
8. Dick, S. J. (1996) *The Biological Universe*. Cambridge University Press, Cambridge.
9. Koestler, A. (1959) *The Sleepwalkers*. Hutchinson, London.
10. Caspar, M. (1948) *Johannes Kepler*, Stuttgart.
11. *Ibid*.
12. Koestler, A. (1959) *The Sleepwalkers*. Hutchinson, London.
13. Rosen, E. (transl.) (1967) *Kepler's Somnium*. University of Wisconsin Press, Madison and London.
14. Christianson, G. (1976) Kepler's *Somnium*: science fiction and the Renaissance scientist. *Science Fiction Studies*, #8, **3**(1).
15. Caspar, M. (1959) *Kepler*. Abelard-Schuman, New York and London.
16. Koestler, A. (1959) *The Sleepwalkers*. Hutchinson, London.
17. Nicolson, M. H. (1948) *Voyages to the Moon*. Macmillan, New York.
18. Brake, M. (2006) On the plurality of inhabited worlds: a brief history of extraterrestrialism. *International Journal of Astrobiology*, **5**(2), 99–108.
19. Parrinder, P. (2000) *Learning from Other Worlds: Estrangement, Cognition, and the Politics of Science Fiction and Utopia*. Liverpool University Press, Liverpool.
20. Dick, S. J. (1996) *The Biological Universe*. Cambridge University Press, Cambridge.
21. Rose, M. (1982) *Alien Encounters: Anatomy of Science Fiction*. Harvard University Press, Cambridge, MA.
22. Wilkins, J. (1640) *The Discovery of a World in the Moone*. London.
23. Godwin, F. (1638) *The Man in the Moone*. London.
24. de Bergerac, C. (1971) The comical history of the Moon. In: *The Other World or the States and Empires of the Moon*. London.

25. Clarke, A. C. (2000) *Greetings, Carbon-Based Bipeds*. Voyager, London.
26. de Bergerac, C. (1971) The comical history of the Moon. In: *The Other World or the States and Empires of the Moon*. London.
27. de Fontenelle, B. (1992) *Conversations on the Plurality of Worlds*. University of California Press, Berkeley, CA.
28. Stimson, D. (1972) *The Gradual Acceptance of the Copernican Theory of the Universe*. Cambridge, MA.
29. Koestler, A. (1959) *The Sleepwalkers*. Hutchinson, London.
30. Jones, R. F. (1951) *The Seventeenth Century*. Stanford University Press, Stanford and London.
31. Jones, R. F. (1953) *The Triumph of the English Language*. Stanford University Press, Stanford and London.
32. Sprat, T. (1667) *The History of the Royal Society of London*. London.
33. Bacon, F. (1857–74) *The Works of Francis Bacon* (eds. J. Spedding, R. L. Ellis and D. D. Heath), 14 vols. Longman, London.
34. White, H. (1968) *Peace Among the Willows*. Martinus Nijhoff, The Hague.
35. Farrington, B. (1951) *Francis Bacon, Philosopher of Industrial Science*. Lawrence and Wishart, London.
36. Bacon, F. (1857–74) *The Works of Francis Bacon* (eds. J. Spedding, R. L. Ellis and D. D. Heath), 14 vols. Longman, London.
37. *Ibid*.
38. *Ibid*.
39. *Ibid*.
40. Crowther, J. G. (1960) *Francis Bacon, The First Statesman of Science*. Cresset Press, London.
41. Orwell, G. (1968) Politics vs. Literature – an examination of *Gulliver's Travels*. In *The Collected Essays, Journalism and Letters of George Orwell*. Secker & Warburg, London.
42. Swift, J. (1726) *A Voyage to the Houyhnhnms*. Book IV of *Gulliver's Travels*. London.
43. Davis, H. (1964) *Jonathan Swift: Essays on His Satires and Other Studies*, Oxford University Press, Oxford.
44. Swift, J. (1726) *Polite Conversations Dialogue III*. Folcroft.
45. Bernal, J. D. (1965) *Science in History*, Vol. II. C. A. Watts & Co., London.
46. Ginsberg, M. (1953) *The Idea of Progress*. Methuen, London.
47. Jones, R. F. (1961) *Ancients and Moderns: A Study of the Rise of the Scientific Movement in Seventeenth Century England*. Washington University Press.

48.   Descartes, R. (1949) *Discourse on Method*. London.

Chapter 2

REMEMBRANCE OF THINGS TO COME: THE MECHANICAL AGE

On 30 June 1860, Darwin's Bulldog strode into a now legendary meeting of the British Association for the Advancement of Science in Oxford. The eager audience had grown hugely. Only minutes before, the meeting had adjourned to the great library of the Museum. Thomas Huxley's opponent for the evening was one Bishop Wilberforce of Oxford. A fervent speaker, nicknamed Soapy Sam for his habit of rubbing his hands as he sermonized, Wilberforce was about to meet his match.

Setting the tone for the battle that was to follow, Wilberforce condemned Darwin's theory as 'a dishonouring view of Nature... absolutely incompatible with the word of God'[1]. Becoming carried away by his oratory, the meeting took a decisive twist. Turning to his antagonist, Wilberforce begged to know, was it through Huxley's grandfather or his grandmother that he claimed descent from a monkey? Huxley slowly and deliberately rose, quiet and grave, whispering 'The Lord hath delivered him into mine hands'[2], and replied:

A man has no reason to be ashamed of having an ape for his grandfather. If there were an ancestor whom I should feel shame in recalling it would rather be a man who plunges into scientific questions with which he has no real acquaintance, only to obscure them by an aimless rhetoric...[3]

The attack on religious prejudice electrified the meeting. One woman fainted and was carried out. Many others jumped to their feet in the excitement. Captain FitzRoy of the famous *Beagle* paced up and down, brandishing the Bible, chanting 'The Book, the Book!'[4]. Once the meeting was over:

> every one was eager to congratulate the hero of the day... some naive person wished it could come over again; and Mr Huxley, with the look on his face of the victor who feels the cost of victory, put us aside saying, 'Once in a life-time is enough, if not too much.'[5]

But the drama of Darwinism had just begun. It was about to be used to justify ruthless exploitation, the conquest of innocent peoples, and even war itself. Only a few condemned these atrocities in the name of science. One of those radical challengers was Huxley's own student, H. G. Wells...

New Atlantis realized

By the nineteenth century Bacon's vision of material progress was realized. Science had secured its dominion over nature. The House of Solomon established its authority in the clanging new workshop of the world that was Victorian Britain. 'Were we required', wrote Thomas Carlyle in 1829, 'to characterise this age of ours by any single epithet, we should call it the Mechanical Age'[6].

Newton's system of the world was set free. The *philosophical* engine, the early steam engine, drove locomotives along their metal tracks; the first steamships crossed the Atlantic; the great transport magnates were building bridges and roads; telegraphs ticked intelligence from station to station; the Lancashire cotton works glowed by gas; and a clamorous arc of iron foundries and coal mines powered this Industrial Revolution.

Newton had created a clockwork cosmos, a mechanical world-view. As the machinery began to mesh, science encroached upon all aspects of life, meeting every challenge with a different invention. Progress and technology seemed inseparable. The machines of science were devised not merely to explore nature, but to exploit it. For every factual gadget, fiction spawned a thousand visions. A selection of key works will reveal some of those visions and the way in which the mood of the age changed.

Science fiction was deeply divided into conforming and radical trends. The Grub Street Presses in London revolutionized the novel. In 1800, speculative fiction was written for the rulers of society. Optimism grew boundlessly. The outlook of the *Voyages Extraordinaires* of Jules Verne identified science with the progressive expansion of capitalism.

Against such cheer stood the strange fruits of science in the scarred landscapes of the industrial districts. Mary Shelley's *Frankenstein* (1818) was a seminal voice of contradiction and dissent, a response to the double-edged sword of technology and change.

As the century wore on, the dark clouds of war and revolution loomed larger. Darwin's *Origin of Species* (1859) infused science with a sense of history. Apocalypse haunted an obsession with a technologically evolving society. Edward Bulwer-Lytton, *The Coming Race* (1871) foretold the race of the future, and satirist Samuel Butler anticipated machine intelligence, setting 'Darwin Among the Machines', in his farsighted *Erewhon* (1872).

By the end of the era, science fiction was written for those who wished to change the world. Evolution became a social force. Darwin and natural selection travelled into space with French astronomer Camille Flammarion, *Récits de l'infini (Stories of Infinity,* 1872*)*, and H. G. Wells explored social Darwinism in *The Time Machine* (1895) and *The War of the Worlds* (1898).

The ghost of electricity

By the dawn of the Mechanical Age, capitalism was exuberant. Yet the first signs of decline had emerged. The belching chimneys of the dark satanic mills found their opposition in the Romantic poets. Romanticism was a secular and intellectual movement. It rebelled against the aristocratic order of the past, and was critical of the rationalization of nature by the New Philosophy.

John Keats talked of unweaving the rainbow[7], suggesting that Newton had destroyed the beauty of nature by splitting and analysing light with a prism. William Blake, another strident opponent of Newton, related the 'philosophy in vogue'[8] to the 'lucre of traffic and merchandise'[9], favouring pastoral idyll over the blight of industry.

A fascination with electricity crackled into life. Founding father Benjamin Franklin had brought electricity down to Earth through the lightning conductor. Michael Faraday conjured a cocktail of electricity and magnetism in the dynamo. The exposed sciatic nerve of a frog's leg kicked Luigi Galvani into the discovery of bioelectricity.

Yet there was darker magic still. As the first science to materialize after Newton, electricity had a long and legendary history. From ancient times people had treasured the doctrine of affinities. The attraction of amber illustrated the very idea of *virtue* dwelling within a special substance. As if through magic, the magnet's enchanting property of virtue was bestowed on other objects through touch alone. The possibilities seemed endless.

Into this climate came Mary Shelley. Her epoch-making *Frankenstein: or, The Modern Prometheus* (1818) is a work of astonishing insight, and from its very genesis a marvellous tale to tell.

In June 1816, Mary Wollstonecraft Godwin and her intended, Percy Bysshe Shelley, had visited the leading figure of Romanticism, Lord Byron, at Lake Geneva. That 'Year Without A Summer' it had seemed the entire planet was frozen in a volcanic winter triggered by the eruption of Mount Tambora the previous year. Along with Byron's physician, John William Polidori, kept indoors by the incessant bad weather, the Romantics turned to conversation. Their reading matter was mostly fantasy, including *Fantasmagoriana*, an anthology of German ghost stories. They talked of animating life. One experiment, credited to Erasmus Darwin, reported that a piece of vermicelli preserved in a glass case had begun to move with voluntary motion.

Byron invited the Shelleys and Polidori to each create an alienating fiction. Mary's tale took form after she fell into a waking nightmare in which a 'pale student of unhallowed arts kneeling beside the thing he had put together' haunted her. The nightmare was the seed of *Frankenstein*. Polidori was motivated by an unfinished story of Byron's to produce *The Vampyre* (1819), the progenitor of the vampire genre. In a single night, the Frankenstein and vampire themes were born.

The story of *Frankenstein* is set in elemental Arctic and Alpine landscapes rather than drab industrial London. The Romantics knew they were living in a new age, signified by Mary's choice of Switzerland for the setting of her novel. The country was reformed only in 1815, the year before the Shelleys visited. Mary considered herself a Modern. She inhabited a Newtonian world, where suddenly everything was being called into question. It was no accident she had Victor Frankenstein study at Ingolstadt University, a prominent centre for science in its time.

Frankenstein is essentially the tale of Victor Frankenstein's alien creature. In stark contrast to the crude distortions of film, stage and television, the novel is complex. Through a series of narrators, including the eloquent 'monster' himself, the book presents a tale of the difference between the old divine order and the new sceptical world of science.

Victor rejects the dark arts of old world alchemists Paracelsus, Albertus Magnus and Cornelius Agrippa, and turns to face the future. He becomes obsessed with the essence of life. He manages to unravel the agency through which dead matter may be given the vital spring of life. Intending his creation to be beautiful, Victor builds a mechanically sound but grotesque creature using cadaver spares from charnel-houses[10]. Only when inspired by the new unbridled science is he gifted this terrible triumph of creation.

Frankenstein: science gone astray

Mary Shelley's *Frankenstein* was spun on the crackling loom of the emergent science of her day. Though a mere 19 years of age when writing her famous novel, Mary had studied Condillac's *Treatise on the Sensations*[11], and read Erasmus Darwin, Humphry Davey,

Figure 2.1 Frontispiece from the 1831 edition of *Frankenstein*.

and Joseph Priestley. Between the first edition in 1818 and the second in 1831, the first volume of Lyell's *Principles of Geology* (1830) was published, dramatically increasing the known age of the Earth. The new geologists were industriously rooting reptilian bones out of the mud; extinct monsters brought back to life. The dinosaur mania for the great fossil lizards and the 'Bone Wars'[12] were about to begin.

It is possible that Mary implies that the creature is spontaneously generated from dung and rot. The notion of electrical force as the agency was later promoted both on stage and film. However, the preface to the 1831 edition of the book refers to the stimulating new science of galvanism. In 1791, Galvani had become the first to grasp the link between electricity and animation. Throughout Europe there was excitement about the application of this new force, and feverish research on the potential of electricity to generate and sustain life.

This potential is the stuff of *Frankenstein*: science gone astray. In the century before Mary Shelley, many writers had hunted for new standpoints from which to examine the human condition of the age. Mary's brilliance was to offer an alternative viewpoint. Whatever process of spontaneous generation raises the creature is immaterial. For Mary it is the consequences rather than the techniques of science that are crucial.

Victor Frankenstein is the Faust of the New Philosophy. The subtitle of the novel, *The Modern Prometheus*, also compares with Prometheus's theft of fire from the gods for profit. Victor's dream is unlimited power through science. Using these archetypes, Mary enhances the mythic power of *Frankenstein* and eclipses the old myths. She shows how Victor's power is brought about by human, not supernatural, agency.

Victor's ambition is met only after he discards the ancient texts to dabble with the new science in his 'workshop of filthy creation'. The new researchers seduce him:

> They ascend into the heavens: they have discovered how the blood circulates, and the nature of the air that we breathe. They have acquired new and almost unlimited powers; they can command the thunders of heaven, mimic the earthquake, and even mock the invisible world with its shadows.[13]

Mary's masterpiece is revolutionary in its portrayal of the tension between man and the new science. God is nowhere to be found. The divine spark of life in Michelangelo's *Creation of Adam* is traded for the galvanized spark of the New Philosophy. Science eclipses religion. With knowledge, however, comes power. There is horror in the new responsibility facing the rationalists. The picture of man's innate simian curiosity in an age of science also makes *Frankenstein* the first evolutionary novel. It foretells our hopes and fears for scientific progress, and the dreadful secrets of the human frame[14].

Frankenstein's creature: loving the alien

Frankenstein's creature looks forward to the monstrous mass of droids that were to advance across the screen of twentieth century cinema. Mechanical Man was one element of the new divided Universe: part physical, part moral. To Descartes, men were mere machines. The business of science was the merely physical. Mary Shelley's novel, however, is less concerned with machinery than with viscera. For Descartes there was a link between mechanical man, his movements governed by physical law, and the rational soul within. And in the creature we do indeed find a soul.

Frankenstein is a diseased creation myth. It is the power of knowledge, and the formation of the alien creature, that lead to our estrangement. The creature's moving account of his life bears witness to an education. Through the strange interior of the creature's mind we see our own conditions of life from a new perspective. He weeps at an account of the many native peoples dispossessed by the invading Europeans. In this way, we see ourselves as the monsters, responsible for vice and bloodshed. We share the creature's disgust with 'civilized' human society. We identify with the alien.

Rather than machine or robot, the idea of Frankenstein's creature leaps into the science fictional future of artificial life. The *nature* of life itself is under the microscope. It is best to think of the creature as a *tabula rasa*[15], having no innate evil other than that learnt through abuse by humans. He is taught to be monstrous. Otherwise, he is an embodiment of the simian in mankind; he need merely be given the opportunity to be a monster.

Once the creature becomes, in his own words, a fallen angel and a malignant devil, he is more often described in monstrous terms. Highlighting the creature's inhumanity, Victor sees him 'descend the mountain with greater speed than the flight of an eagle, and quickly lost him among the undulations of the sea of ice'[16]. Understandably, Victor stops short of creating a race of such devils.

The taste and feel of science

Frankenstein calls for vigilance in the practice of science. It warns of the primal urges of power and control in all creations of technology. The book became a potent metaphor of the powerlessness of the inventor. Ever since *Frankenstein*, codified in such artefacts as the hydrogen bomb and human genetics, it has proved difficult to limit the social fallout of knowledge.

As with Kepler and Swift, Mary Shelley's field of interest was the conflict between the human and the nonhuman[17]. It is unsurprising that she was part of the Romantic Movement. Most mainstream fiction since the Renaissance has been unconcerned with the nonhuman world revealed by science. Poetry had little to do with the laws of physics.

For the Romantics, however, and for science fiction, a dialogue with the nonhuman is the key concern. Witness Wordsworth's interest in science, which hints at the science fiction of the future:

> If the labours of men of science should ever create any material revolution... in our condition... the poet will sleep then no more than at present, but he will be ready to follow the steps of the man of science, not only in those general indirect effects, but he will be at his side, carrying sensation into the midst of the objects of the science itself.[18]

Trying to best express 'the taste, the feel, the human meaning of scientific discoveries'[19] is how science fiction works. It has been a provocative and compelling touchstone of the dialectic of science and progress. It has presented a mode of thinking whose discourse is the reducible gap between the new worlds uncovered by science and the fantastic strange worlds of the imagination.

Those dreadful hammers

Great engines of change were turning over the soil of the world. The fossil record churned out signatures of beasts no longer to be found, challenging biblically literal accounts of natural history. Some would not be swayed by such evidence. Thomas Jefferson urged pioneers heading west to search for the woolly mammoth. A deluded evangelical naturalist even reported having heard one trumpeting through the dark forests of Virginia[20].

Christian scholars had earlier estimated the age of the world by adding up the 'begat's. In this way, Newton had concluded, given the years elapsed between Adam and Abraham, that the date of Creation was a mere 3998 BC; Kepler had dated it at 3993 BC. This curious approach was taken to its most bizarre conclusion by the Bishop of Armagh, James Ussher, who had calculated 'the beginning of time... fell on the beginning of the night which preceded the 23rd of October, in the year... 4004 BC'[21]. Such thoroughness is comforting.

As the death roll of extinction grew, the French zoologist George Cuvier founded the science of palaeontology. The new geology revolutionized the Mechanical Age. Its effect spread far beyond the scientific horizon, destroying established truths, forcing all to confront the terrible extent of history. John Ruskin was moved to comment, in 1851, 'If only the Geologists would let me alone, I could do very well, but those dreadful Hammers! I hear the clink of them at the end of every cadence of the Bible verses'[22].

Jules Verne's *Journey to the Centre of the Earth* (1864) wielded the new geology like a club. If Charles Lyell's *Principles of Geology* was the manifesto of the emergent science, Verne's novel was its electioneer. A voyage narrative through a subterranean world, Verne's book explodes the belief that the Earth had been in stasis since the Creation, a mere six thousand years ago.

Verne's creative journey had begun in 1863 with the first of 63 *Voyages Extraordinaires: Voyages in Known and Unknown Worlds*. An early advert claimed Verne's goal was 'to outline all the geographical, geological, physical, and astronomical knowledge amassed by modern science and to recount, in an entertaining and picturesque format that is his own, the history of the universe'[23]. Some mission.

Verne was a technophile. In contrast to mainstream French science fiction, his efforts established the marvellous machine as a most enduring motif. His novels, populated with aircraft, space-ships, and submarines, inspired one commentator to suggest that 'it is the machine that is the hero'[24] in Verne's work. As the century clanked on, rather than fearing the catastrophic machine, Verne's technophilia prevailed[25].

Subterranean fire

Journey to the Centre of the Earth is classic Verne. The new science had its first great sceptic in Shelley, and its chief positivist in Verne. Though the mythology of the machine is absent, the penetrative thrust of empire is unmistakable. Reassuringly free of any offensive on bourgeois society, the book promotes a heady confidence in progress. Verne's is a predictable Universe in which the unknown is easily assimilated into our taxonomies.

The feeling of estrangement in science fiction is bound to the scientific worldview, and the alienating discovery of the new Universe. Though this separation from nature began with Copernicanism, it reached its peak in the Mechanical Age. The sheer pace of dizzying change was a key factor. So too was the Victorian crisis in faith[26] hastened by the emergent sciences of biology and geology. The modern age of alienation had truly begun. The science fiction of the age can be seen as an attempt to repair this sudden separation from nature, to reload the emptiness, to somehow jack-in to the void.

Verne's book is dominated by such a conquest of space. Gone are Dante's mythical speculations of an Earthly core, locus of the Devil and his legions. In its place is the quest to possess nature absolutely for science. To reach the core of the world is to achieve completion, to pierce the living heart of nature, the glittering prize.

Axel, nephew of a distinguished German geologist, Professor Lidenbrock, tells the story of the journey. Pouring over a mediæval volume written by the Icelandic alchemist Arne

Saknussemm, the Professor had discovered a runic cryptogram. The code suggests entry to the Earth's interior can be gained through the cone of a defunct Icelandic volcano, named Sneffels: 'Descend, bold traveller, into the crater of the jokul of Sneffels, which the shadow of Scartaris touches before the kalends of July, and you will attain the centre of the earth; I have done this, Arne Saknussemm'.[27]

The decoding of this cipher is Verne's opening salvo, written at a time when Christian chronologists clashed with the age-dating geologists. The stark contrast was that the geologists studied not Scriptures, but stones[28]. This is how Verne's compatriot, the naturalist George Louis Leclerc, had communicated the geologist's creed in 1778:

> Just as in civil history we... decipher ancient inscriptions... so in natural history one must dig through the archives of the world, extract ancient relics from the bowels of the earth, [and] gather together their fragments.... This is the only way of fixing certain points in the immensity of space, and of placing a number of milestones on the eternal path of time[29].

Scottish geologist James Hutton had breathtakingly declared in 1785 that the rocks revealed 'no vestige of a beginning, no prospect of an end'. English engineer William Smith had been among the first to decipher the hieroglyphics of the stones, once the steam engine had opened up the veins of the world. Similarly, Verne presents his novel's paradigm: nature is a cipher to be cracked. The key to the runes discovered by the Professor is that, much like the strata themselves, they have to be read backwards. Indeed, Verne's story derives its force from the notion that the expedition is also a quest into the depths of evolutionary time.

The trail blazed by geologists such as Hutton and Leclerc had also inspired the palaeontologists. George Cuvier had anticipated the idea of species extinction. Once in the subterranean caverns, grottos and waters, Axel and the Professor find the interior alive with prehistoric plant and animal life. A herd of mastodons, giant insects, and a deadly battle between an ichthyosaurus and a plesiosaurus follow.

A giant prehistoric man found overlooking the mastodon herd is another of Verne's nods to contemporary science. When the Professor lectures on the latest anthropological discoveries, he refers to Boucher de Perthes, who in 1863 had unearthed a human jaw in northern France, suggesting that man was over 100,000 years old. Verne waited until the discovery was confirmed before including it in his 1864 novel. Significantly, this panorama is subjected to an orgy of classification at the hands of the travellers, Axel and the Professor. To name is to appropriate and conquer. Their taxonomy is an attempt to bleed nature of its strangeness, to render it human[30].

From the outset Professor Lidenbrock is portrayed as an avatar of science, locked in a lethal struggle with the nonhuman world. The Professor's Baconian pursuit to triumph over nature is contrasted with Axel's romanticism. Witness a typical dialogue when the travellers discover that a cavernous sea is subject to tidal forces such as those on the Earth's surface. Axel is thrilled and enchanted. The Professor, a model of bourgeois rationalism, merely lectures that subterranean waters are as prone to gravitation as any other sea:

'Here is the tide rising,' I cried.

Yes, Axel; and judging by these ridges of foam, you may observe that the sea will rise about twelve feet.'

'This is wonderful,' I said.

'No; it is quite natural.'

'You may say so, uncle; but to me it is most extraordinary, and I can hardly believe my eyes. Who would ever have imagined, under this terrestrial crust, an ocean with ebbing and flowing tides, with winds and storms?'

'Well,' replied my uncle, 'is there any scientific reason against it?'

'No; I see none, as soon as the theory of central heat is given up.' 'So then, thus far,' he answered, 'the theory of Sir Humphry Davy is confirmed.'[31]

At the end of the book, Axel, considering the 'wonderful hypotheses of paleontology'[32], is finally won over to the Professor's viewpoint. Though they never do reach the Earth's core, the narrative takes one final twist. On reaching the Earth's surface there is the frustrating mystery of the travellers' compass. The wayward compass indicates that the explorers had travelled 1500 miles off course. The Professor is apoplectic, 'for a scientist an unexplained phenomenon is a torture of the mind'[33]. However, recent convert Axel realizes that the compass poles had merely been reversed during a subterranean electrical storm. The postscript confirms our faith in the methods of science. To the last word, nature is a code to be broken.

The incendiary evolutionists

The man who travelled furthest in time started his career a Creationist. Charles Darwin, however, was gifted with a familial vision that proved hard to live down. His grandfather Erasmus Darwin was an ingenious mechanic, inventing a speaking machine, a mechanical ferry and a rocket motor long before the dreams of Russian rocket pioneer Tsiolkovsky[34]. He was also a provocative evolutionist. As a boy Charles had pored over Erasmus's mighty work on evolution, *Zoomania*, published in two volumes in 1794 and 1796. It was replete with hearty exclamations that life had evolved from a single ancestor.

Erasmus Darwin was a celebrated communicator of science. Romantic poet Samuel Taylor Coleridge declared him 'the first literary character of Europe, and the most original-minded man'[35]. One of Erasmus's poems on evolution enjoys a science fictional vision. It foresees, with unerring accuracy, a future of colossal skyscraper cities, overpopulation, convoys of nuclear submarines, and the advent of the car[36].

Charles did not inherit his grandfather's boldness of spirit. Science historian A. N. Whitehead proclaimed that 'Darwin is truly great, but he is the dullest great man I can think of'[37]. Constrained by tradition and Church, like Copernicus before him, Darwin kept his earth-shattering theory concealed for twenty years. The blow to Christianity and the social order struck by his theory would, he dreaded, inspire 'atheistic agitators and social revolutionaries'[38]. Evolution should have been countenanced long before. Opposition from landed and clerical interests, however, feared its deadly threat to the divine ordering of the world.

Darwin struck at the heart of humanity itself. Newton's system of the world had essentially re-established the integrity of design, which had been shattered by Copernicus and Galileo. The Christian picture of creation had stayed more or less untouched. Man was still made in the image of God. After Darwin, the book of Genesis lay in shreds as a literal history.

The Origin of Species (1859) was appropriated by the radical, anti-clerical wing in politics, and twisted to its agenda of *laissez-faire* capitalism[39]. Darwin unwittingly provided an alibi for brutal exploitation by the 'fittest'; the subjugation of lesser by higher peoples. Association with nature 'red in tooth and claw', as Tennyson put it, could justify even war itself. The notion of the Chosen Ones, that one-time apology for the supremacy of classes or races, had withered. It was replaced by a 'Darwinian' validation of a brave new world of reason, industry and empire.

Darwinism injected the lifeblood of history into science. 'He who... does not admit how vast have been the past periods of time may at once close this volume', Darwin wrote in the *Origin*. The theory of evolution might have been used to unite the human and nonhuman spheres. Instead of such an emphasis on the affinity of man and nature, the social evolution of humanity was eclipsed by scientism. A limited science focus resulted in the perverse

Figure 2.2 A typical satire: an 1874 caricature of Darwin from the *London Sketchbook*.

justification of race theories and imperialism. It was the utopian tale of the future that explored the connection between nature and society.

The power of the coming race

Darwin's was the definitive philosophy in the Mechanical Age. It transformed all spheres of thought – scientific, social, political, spiritual, and artistic. The utopian tale provided a vehicle for concerns about an increasingly urbanized humanity. It also examined the social implications of evolution. The irresistible rise of the metaphor of evolution spawned around seventy futuristic fantasies in England alone between 1870 and 1900.

The idyllic vision of a static world became passé. In its place was mutability. The utopian tales stressed the ebb and flow of evolution, as a reaction to the unsettling changes in the fabric of Victorian society. After Darwin, the new paradigm was the process of becoming; the question as to what would become of man.

In the words of Tennyson 'Earth's pale history runs, What is it all, but a trouble of ants in the gleam of a million million suns?'[40]

Progress gave way to pessimism. The economic slump of the Great Depression in Britain between 1873 and 1896 marked the end of unquestioned expansion. In 1870 the country had been Liberal, but by the middle of the 1890s British politics was sharply polarized. Most of its capitalists had seceded to the Conservatives. The first cloth-capped proletarian socialist sat in Parliament[41]. The utopian fiction arose as the social order began to corrode. Many of the narratives offered a parody on the disorder of the contemporary culture[42].

Science was not wholly responsible for the Great Depression and its ills. It was correctly felt, however, that science had transformed industry. This revolution had cultivated the urban, class-conscious culture of industrial Britain. In addition, the radical displacement of Christian faith brought about by Darwin had ironically led to an alienation from nature. As the problems of the Mechanical Age grew more complex and challenging, historians of the future delivered bleaker forecasts.

Edward Bulwer-Lytton's *The Coming Race* (1871) is a book about supermen. Edward George Earl Bulwer-Lytton, 1st Baron Lytton, politician and novelist, was friend to the creator of the modern British Conservative Party, Benjamin Disraeli. It was Lytton's intention to satirize both Darwinian biology and the ideals of John Stuart Mill's *The Emancipation of Women*. His novel includes a pseudo-scientific account of an evolved line of humans who believe they are descended not from apes but from frogs. The book also contains a somewhat inelegant gender reversal. The women are fitter, beefier, more assertive and hairier than the men.

His fascinating, if bizarre, tale is set in a subterranean world of well-lit caverns. It begins as the narrator, an American mining engineer, falls into an underground hollow. There he discovers a mysterious human-like race, the Vril-ya. These humanoids derive immense power from *vril*, an electromagnetic animating force which fuels air boats, mechanical wings, formidable weapons and automata:

> In all service, whether in or out of doors, they make great use of automaton figures, which are so ingenious, and so pliant to the operations of vril, that they actually seem gifted with reason. It was scarcely possible to distinguish the figures I beheld, apparently guiding or superintending the rapid movements of vast engines, from human forms endowed with thought[43]

Bacon's dream has been realized by the subterraneans. The unearthing of *vril*, the 'all permeating fluid', borne by strident emancipated females, has enabled the race to master nature. Gender equality has been achieved, and war eliminated through mutually assured destruction. It is a utopia made real 'the dreams of our most sanguine philanthropists'[44].

Lytton, however, rejects this 'angelical' social order. The sociable community of the Vril-ya has eliminated competition, but is barren of those 'individual examples of human greatness, which adorn the annals of the upper world'[45]. Conflict and competition, misery and madness, are all innately human. Sounding a note struck more clearly in Huxley's *Brave New World*, Lytton brands the aim of 'calm and innocent felicity' as a vain dream. Utopia, and the displacement of human industry to *vril*, would lead only to enervation and ennui[46].

The Coming Race marks the start of the Victorian obsession with evolving society. A new kind of life has been secured through the application of science. The novel suffers from machine determinism, however. The new technology has no social agency, even though it is described as having inevitable social consequences. Far-flung subterraneans who do not have *vril* are uncivilized (or, we might say, 'the great unwashed', since the aristocrat Lytton is alleged to have coined the phrase). Indeed, the possession of *vril* energy *is* the civilization, and the refinement of society is based on technology alone.

The book strikes one final fearful note. As suggested by the ominous title, once the more advanced Vril-ya surface from their caverns, they will take the place of man: 'the more deeply I pray that ages may yet elapse before there emerge into sunlight our inevitable destroyers'[47].

Meanwhile, back in the Mechanical Age, John Lawson Johnston found his own utopia. Inspired by Lytton he made a fortune from a strength-giving beef extract elixir – known as Bovril.

Darwin among the machines

The harnessing of technology made the Vril-ya formidable. It is the rejection of the machine, however, that gives strength to the Erewhonians in Samuel Butler's irreverent satire *Erewhon* (1872). Resolving to make his fortune in a foreign land, a traveller uncovers the beautiful faraway realm of Erewhon. The attractive Erewhonians provide a home for the visitor, who soon learns that this idyll has its flaws.

The key to the novel is Butler's dialectical approach. He presents both sides of arguments on Victorian life and thought. In *Erewhon*, criminals are indulged as if they are sick. The ill and the poor are brutally punished. As with Swift, *Erewhon* skilfully conjures up an ambiguous utopia.

The most Darwinian part of *Erewhon* is the 'Book of the Machines'. It revolves around the question of whether the machine is servant of man, or man servant to the machine. Butler cites evidence of the growing dominion of machine over man:

Consider also the colliers and pitmen and coal merchants and coal trains, and the men who drive them, and the ships that carry coals – what an army of servants do the machines thus employ! Are there not probably more men engaged in tending machinery than in tending men?[48]

To further his machine hypothesis, Butler uses material from his earlier essay, *Darwin Among the Machines*. He imagines a future of evolved machine consciousness, and sees the downfall of mankind:

> There is no security against the ultimate development of mechanical consciousness.... Assume... that conscious beings have existed for some twenty million years: See what strides machines have made in the last thousand! May not the world last twenty million years longer? If so, what will they not in the end become? Is it not safer to nip the mischief in the bud and to forbid them further progress?[49]

So he has the Erewhonians ban all machines, consigning them to the culture's museums.

Butler had doubts about Darwin. On the one hand he was a staunch evolutionist. Darwin was a great redeemer from the fetters of biblical literalism. Accused of reducing 'Mr. Darwin's theories to an absurdity'[50], Butler replied, 'nothing could be further from my intention'[51].

Butler's purpose was to mock those who argued that the world had been designed for man. He objected to the contradictory concept of evolution as mechanical and at the same time unsolicited. If evolution is the result of accidental changes mechanically perpetuated, then it is best left to describing the progress of machines. A theory that regards man as a machine is no better than an absurdity that supposes machines to be animate.

With ironic chance insight, Butler invented machine consciousness. *Erewhon* is the prototype of a twentieth century obsession. Butler can never have dreamt of the thousands of artificial intelligences that would follow. Ever since Butler, science fiction has had double vision on the metaphor of the machine. At times, machines mediate between man and the Universe, human and nonhuman, acting as the agency of man's protection. On other occasions, machines act as a medium for the nonhuman, a threat to the human condition.

Witness *The Time Machine* (1895), where the metaphor flows both ways, and H. G. Wells masters the machine both as a symbol for the power of reason, and as a diabolical mechanism.

The Time Machine

Herbert George Wells emerged from an English lower middle class which had previously spawned only one other key author – Charles Dickens. Wells' mother had been in service, his father a gardener. Though they were hopeful of elevating their status on becoming shopkeepers, the shop failed, year after year. Wells' own employment began as a draper's apprentice. It ended rather abruptly when he was told he was not refined enough to be a draper. Such rejection at the sharp end of a class conscious Mechanical Age became the motivation for Wells' critique of the world.

Wells' watershed came on meeting Darwin's Bulldog. Wells had won a scholarship to the Normal School of Science, later the Royal College of Science, studying evolutionary biology under the great T. H. Huxley. A fervent Darwinian, Huxley was the chief science communicator of the Mechanical Age. He had created the phrase 'agnostic', and impressed man's hominid ancestry on the public imagination. His public lectures attracted huge audiences. Two thousand were reportedly turned away at St Martin's Hall in 1866[52], the year of Wells' birth.

Huxley's crowning triumph had come from the infamous conflict with Bishop Wilberforce. The feud finally ended when Wilberforce dashed his head while horse riding, making him fortunately oblivious to Huxley's last judgment, 'for once reality and his brain came into contact, and the result was fatal'[53].

With Huxley as his inspiration, Wells began as an author, living in the dark, lanterned, black macadam streets of Victorian London, engine-room of the British Empire. The first of Wells' seminal novels, *The Time Machine*, plotted a dark future for man. The book was a sceptical view of the devilish enginery of progress and imperialism. It was an instant triumph.

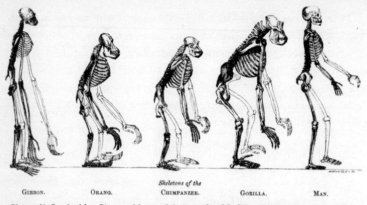

GIBBON.   ORANG.   *Skeletons of the* CHIMPANZEE.   GORILLA.   MAN.

Photographically reduced from Diagrams of the natural size (except that of the Gibbon, which was twice as large as nature),
drawn by Mr. Waterhouse Hawkins from specimens in the Museum of the Royal College of Surgeons.

Figure 2.3 The frontispiece from Huxley's 1863 *Evidence as to Man's Place in Nature*, the first airing of the famous image comparing the skeletons of apes to Man.

The Time Machine has two major themes: evolution and social class. The book is an ingenious voyage of discovery through the invention of a machine, which symbolizes the power of science and reason. The Time Traveller sets out to navigate and dominate time. His discovery: time is lord of all. The significance of the story's title becomes clear. Man is trapped by the mechanism of time, and bound by a history that leads to his inevitable extinction.

The Traveller's headlong fall into the future begins at home. The entire voyage through the evolved worlds of man shows little spatial shift. The terror of each age unravels in the vicinity of the Traveller's laboratory. 'It is not what man has been, but what he *will* be, that should interest us', Wells had written in his essay, *The Man of the Year Million*[54]. In *The Time Machine* we have Wells' answer – a vision calculated to 'run counter to the placid assumption... that Evolution was a pro-human force making things better and better for mankind'[55].

Time's arrow thrusts the story forward to the year AD 802701. The Traveller meets the Eloi, a race of effete, androgynous and child-like humans living an apparently pastoral life. Man's conquest of nature, it seems, has led to decadence. On discovering the subterranean machine world of the albino, ape-like Morlocks, a new theory emerges. Over time, the gulf between the classes has produced separate species:

> At first, proceeding from the problems of our own age, it seems clear as daylight to me that the gradual widening of the present merely temporary and social difference between the Capitalist and the Labourer, was the key to the whole position. No doubt it will seem grotesque enough to you – and wildly incredible! – and yet even now there are existing circumstances to point that way[56]

Initially believing the Eloi to be descendants of the ruling class, the Traveller discovers that the working class have evolved into the bestial Morlocks. These cannibal hominids man the machinery that keeps the Eloi – their flocks – passive:

> The great triumph of Humanity I had dreamed of took a different shape in my mind. It had been no such triumph of moral education and general cooperation.... Instead, I saw a real aristocracy, armed with perfected science and working to a logical conclusion the industrial system of today. Its triumph had not been simply a triumph over nature, but a triumph over nature and the fellow-man[57]

Wells foresaw a bifocal future. One image in the lens, 'upper-world man had drifted towards his feeble prettiness'[58], focuses on what man may become when natural selection is eradicated, as with the Eloi. The lens of the Morlock future, 'the under-world [of] mere mechanical industry'[59] arises when industrialization serves as a chronic condition for natural selection. Wells' warning is all the more powerful for making the reader feel responsible. It is the inequity of contemporary class society that leads to such monstrous futures.

Wells took a momentous leap in the portrayal of evolution. 'People unfamiliar with such speculations as those of the younger Darwin, forget that the planets must ultimately fall back one by one into the parent body'[60]. Evolution is revealed not merely as a biological and social process. It is a cosmic development, played out against a backdrop of dying planets and dying Sun. The vision is one of man being swept away 'into the darkness from which his universe arose'[61].

Rescuing his machine from the Morlocks, the Traveller journeys to the far future. The time machine reappears on a terminal beach. The Earth is locked by tidal forces. The planets spiral toward a red giant Sun, which hangs motionless in an endless sunset. The solar system is in meltdown. In this entropic decay of the cosmic machine, man has become a strange round black creature that hops about 'against the weltering blood-red water'[62]. So ends Wells' terrible account of progressive devolution[63].

Martian machines

The Time Machine reveals Wells' dread of proletarian uprising in its antipathy towards the Morlocks. He eschews a Marxist emphasis on class hostility by changing class difference into species difference[64]. Indeed, the mood of the late Mechanical Age was one of fear. Between 1870 and the start of the First World War there were hundreds of authors writing invasion literature, often topping the bestseller lists in Germany, France, Britain and the United States. This pervading sense of *fin de siècle*, and *fin de globe*, was fed limitless power by Darwinian evolution. It enabled Wells to produce work whose fascination has not dimmed since.

A host of spectres haunted the age, and into this fragile climate marched an alien invasion. In the greatest introduction of science fiction, Wells' *The War of the Worlds* (1898) does for space what *The Time Machine* does for time:

No one would have believed in the last years of the nineteenth century that this world was being watched keenly and closely by intelligences greater than man's and yet as mortal as his own.... With infinite complacency men went to and fro over this globe about their little affairs, serene in their assurance of their empire over matter[65]

Wells' imaginative lens now becomes a telescope; the invading Martians are the 'men' of the future. It is the wrong end of the telescope, however. Imperial Britain is on the receiving end of interplanetary social Darwinism:

... Yet across the gulf of space, minds that are to our minds as ours are to those of the beasts that perish, intellects vast and cool and unsympathetic, regarded this Earth with envious eyes, and slowly and surely drew their plans against us.[66]

The Martians are agents of the void. In Verne's *Journey to the Centre of the Earth*, man had penetrated space. In *The War of the Worlds*, it is space that powerfully shatters the human sphere. The book is the finest and most influential of alien contact narratives. It is the first Darwinian fable on a universal scale.

Wells' wrath is focused on the idea of the becoming of man. The book begins with a quote from Kepler:

But who shall dwell in these worlds if they be inhabited?... Are we, or they, Lords of the World?... And how are all things made for man?[67]

The narrator of this struggle for survival is a philosopher, writing a thesis on the progression of moral ideas with civilization. His conclusion of a bright future is abruptly blown apart in mid-sentence by the brutal natural force of evolution in the shape of the Martian attack.

The War of the Worlds features the first 'menace from space'. Despite Kepler's lunar serpents, the modern alien owes everything to Wells. With their distinctive physiology and intellect the book's Martians are the prototypical alien. Wells' story was first serialized in 1897, the same year that Kurd Laßwitz, father of German science fiction, published his more peaceable view of Martians in *Auf Zwei Planeten* (*On Two Planets*).

Wells was keenly aware of the possibility of life on other worlds. He had contributed to the extraterrestrial debate in 1888 at the Royal College of Science on the topic *Are the Planets Habitable?* Wells had also written essays in support of fellow pluralists Kepler, Camille Flammarion and Percival Lowell, whose mistaken obsession with life on Mars had recently reached Europe. Rather than being a capricious work of fiction, *The War of the Worlds* destroys the idea that man is the pinnacle of evolution. Instead Wells creates the myth of a technologically superior alien intelligence.

Mars is a dying world. Its seas are evaporating, its atmosphere dispersing. The entire planet is cooling, so 'to carry warfare sunward is, indeed, their only escape from the destruction that generation after generation creeps upon them'[68]. So the terror of the void is brought to Earth. Wells' issues repeated reminders of 'the immensity of vacancy in which the dust of the material universe swims'[69] and invokes the 'unfathomable darkness'[70] of space. Life is portrayed as precious and frail in a cosmos that is largely deserted[71].

The book carefully conveys the quality of the void – immenseness, coldness, and indifference – in its rendering of the aliens. It is the Martian machines, however, that vividly hammer home the cosmic chain of command:

> It is remarkable that the long leverages of their machines are in most cases actuated by a sort of sham musculature.... Such quasi-muscles abounded in the crablike handling-machine.... It seemed infinitely more alive than the actual Martians lying beyond it in the sunset light, panting, stirring

Figure 2.4 The Martian tripods, illustrated by Alvim Corréa, 1906.

ineffectual tentacles, and moving feebly after their vast journey across space[72].

The Tripods tower over men physically, as the vast intellects of their occupants tower over human intelligence. Bodily frail, but mentally intense, the Martians and their superior machines are instruments of human oppression. Their weapons of heat rays and poison gas are dehumanizing devices of mass murder. All attempts at contact are futile, furthering the idea of the aliens as an unrelenting force of the void.

Fittingly, the human response to this cosmic struggle is alienation[73]. The narrator dissociates himself from the grim reality of the inevitable triumph of death over life:

I suffer from the strangest sense of detachment from myself and the world about me; I seem to watch it all from the outside, from somewhere inconceivably remote, out of time, out of space, out of the stress and tragedy of it all[74].

Wells began the invasion on bicycle. It is intriguing to picture him mapping mayhem as he 'wheeled about the district marking down suitable places and people for destruction by my Martians'[75]. As early as 1896 he declared his intentions to 'completely destroy Woking – killing my neighbours in painful and eccentric ways – then proceed via Kingston and Richmond to London, which I sack, selecting South Kensington for feats of particular atrocity'[76]. It is the exquisite violence of Wells' imagination that marks his genius[77].

Indeed, it is in South Kensington that the narrator is haunted by the sound of the Martians howling 'Ulla, ulla, ulla, ulla'. The alien invaders have finally fallen prey – to earthly microbes. Their fate emphasizes not only the insignificance of human resistance to the struggle, but also the latent power of unsolicited natural selection.

The story ends with the dialectic of the Martians. On one hand, the narrator feels for the aliens. In the tragedy of the Martians there is the tragedy of man[78]. The aliens' ordeal hints at the common struggle for all life in a hostile Universe. On the other hand, Wells has repeatedly described the Martians as vast, cool and unsympathetic; alien in tooth and claw, as it were. The narrator may merely be projecting emotion onto creatures that are fundamentally inhuman.

Wells' Martians are a fascination. They are alien, yet they are human. They are what we may one day become, with their 'hypertrophied brains and atrophied bodies'[79], the tyranny of intellect alone. They are also political. Wells evidently has the Martians brutally colonize Earth, but 'before we judge them too harshly we must remember what ruthless and utter destruction our own species has wrought... upon its own inferior races.... Are we such apostles of mercy as to complain if the Martians warred in the same spirit?'[80].

Finally, in writing of the alien, Wells is also writing about his own world. The Martians are a barely veiled criticism of the Mechanical Age[81], with its application of science to industry. In this way, *The War of the Worlds* attacks the social machine of capitalism, the reducing of humans to anonymous cattle, the indifference at any attempts to communicate the inhumanity of the system, and the feeling of alienation.

Life in the Universe

The War of the Worlds was an immediate success. One bourgeois commentator in the *Daily News*, however, considered it to be so brutal that it 'caused insufferable distress to the feelings'[82]. The impact of Wells' book has been immense. It has inspired thousands of imitations. *The War of the Worlds* signals the origin of the modern alien in fiction, and its potency in the public imagination.

Yet Wells' was not the only commanding voice. The alien also infused the work of Camille Flammarion, whose fact and fiction infected the popular culture of France[83]. Flammarion's calling as a science communicator began when he was a mere twenty years of age. *La Pluralité des Mondes Habitées* (*The Plurality of Habitable Worlds*, 1862) was a successful study of the habitability of the planets of the solar system. It went through 36 editions by 1892[84], and was translated into fifteen languages[85].

Flammarion's projects proved to be enduring. He founded the Societé Astronomique de France and a Flammarion publishing firm, both of which still exist. He constructed a telescope, gifted by an admirer of his work, twenty miles south of Paris at the Juvisy-sur-Orge estate, which is still operational[86].

A speculative mélange of science and fiction, *Les Mondes Imaginaires et Les Mondes Réels* (*Real and Imaginary Worlds*, 1864), was Flammarion's second book. It is a critical history of extraterrestrial life, which pays tribute to the inspiring cosmic voyages[87] of Kepler, Godwin, Cyrano and Swift, and their speculations on the habitability of planets and stars[88].

Un missionnaire du moyen âge raconte qu'il avait trouvé le point
où le ciel et la Terre se touchent...

Figure 2.5 The enigmatic 'Flammarion Woodcut' originates with Flammarion's 1888 L'atmosphère: météorologie populaire.

Like Wells, Flammarion was passionate about communicating science to ordinary people. It was only a matter of time before he used science fiction to further the cause. His 1872 *Recits de L'Infini*, translated in 1873 as *Stories of Infinity*, comprised three fascinatingly imaginative tales of other worlds. They tell of a disembodied travelling spirit which observes the range of wondrously exotic life in the Universe[89]. With *In Infinity*, *The History of a Comet*, and particularly *Lumen*, Flammarion led the way for the public acceptance of the cosmic perspective of modern science.

Lumen signals the way in which the imagination can inspire thought experiments[90] in science. Written before Verne and

Wells had developed science fiction, *Lumen* is based on the dialogic style of science communication[91]. In *Lumen*, Flammarion was the first to apply a theory of evolution, albeit Lamarckian, to the creation of alien life forms. He laid one of the keystones of twentieth century science fiction:

> We have grown so used to the idea of alien beings since H. G. Wells found a melodramatic role for them to play in *The War of the Worlds* that it is hard to imagine a time when the idea was new and wonderfully exotic[92].

Kepler deserves the credit for inventing the alien in *Somnium*. Flammarion, however, was the first writer to 'extrapolate that notion to its hypothetical limit'[93] and to 'fill that range with examples by the dozen'[94]. Wells' masterstroke was to reflect the inhumanity of the alien back on humanity itself, 'To me it is quite credible that the Martians may be descended from beings not unlike ourselves, by a gradual development of brains and hands... at the expense of the rest of the body'[95]. In the immediate future it was the Wells' image of the monstrous alien that would prevail. Only later did Flammarion's idea of alien life as the precious fifth element reappear.

Space–time

Not content with creating the benign alien of *Lumen*, Flammarion establishes space–time. Thirty years before Einstein's Theory of Special Relativity, *Lumen* was the earliest science fiction novel to suggest that time and space were not absolute. They exist, said Flammarion, only relative to one another[96]. He goes on to consider how travelling faster than light would render history in reverse. Indeed, his notion of space as a seething sea of 'undulations', replete with latent energies, is equally inspired.

Time travel was invented in the Mechanical Age. This invention was bound up with the concept of time. The ancient Greeks had two words for time, *chronos* and *kairos*. While *kairos* is qualitative, *chronos* is quantitative in nature. Industrialization involved a mechanistic approach to time, which appears to have inspired the collective imagination. This quantitative approach found its way into fiction in Mark Twain's *A Connecticut Yankee in King Arthur's Court* (1889) and in Lewis Carroll's *Sylvie and Bruno* (1889), where the hands of a professor's Outlandish Watch can be moved backwards, enabling the holder of the watch to move back in time too.

Wells, however, was the true pioneer of time travel[97]. His 1888 story *The Chronic Argonauts* is an exciting foretaste to *The Time Machine*. The tale features the jauntily named Dr Moses Nebogipfel who journeys from 1887 to 1862, where he kills someone in self-defence and returns in time[98]. Dr Nebogipfel speaks of 'a geometry of four dimensions – length, breadth, thickness, and duration' as the basis for time travel[99], heralding the Time Traveller's hypothesis in *The Time Machine*: 'There is no difference between Time and any of the three dimensions of Space except that our consciousness moves along it'[100].

It is this Wellsian notion of time as the fourth dimension that we find in Einstein's sky. It is the fabric of a Universe of effortlessly wheeling galaxies in gently curving space–time. It is into this bible-black sky that science fiction now plunged, trading the Wellsian terror of the void for a Vernian freedom of infinite space–time. 'All philosophy', Fontenelle had written, 'is based on only two things: curiosity and poor eyesight; if you had better eyesight you could see perfectly well whether or not these stars are solar systems'[101]. In the Astounding Age, curiosity and the outward urge took science fiction to infinity, and beyond.

References

1. Huxley, Julian (1903) *Life and Letters of Thomas Henry Huxley*. MacMillan, London.
2. *Ibid*.
3. *Ibid*.
4. De Beer, G. (1964) *Charles Darwin, Evolution by Natural Selection*. New York.

5. Sidgwick, Isabella (1898) A grandmother's tales. *Macmillan's Magazine*, London.

6. Carlyle, T. (1904) Signs of the times. In: *Critical and Miscellaneous Essays*. New York.

7. Keats, John (1818) *Lamia*. London.

8. Ackroyd, P. (1995) *Blake*. Sinclair-Stevenson, London.

9. *Ibid*.

10. A charnel house (Med. Lat. *carnarium*) was a place for depositing the bones which might be thrown up in digging graves.

11. Vasbinder, S. H. (1984) *Scientific Attitudes in Mary Shelley's Frankenstein*. University of Michigan Research Press, Ann Arbor.

12. Dinosaur mania was exemplified by the fierce rivalry between Edward Drinker Cope and Othniel Charles Marsh, both of whom raced to be the first to find new dinosaurs in what came to be known as the 'Bone Wars'.

13. Shelley, Mary (1818) *Frankenstein: or, The Modern Prometheus*. London.

14. Aldiss, B. (1986) *Trillion Year Spree*. Victor Gollancz, London.

15. Clute, J. and Nicholls, P. (1999) *The Encyclopaedia of Science Fiction*. Orbit, London.

16. Shelley, Mary (1818) *Frankenstein: or, The Modern Prometheus*. London.

17. Rose, M. (1982) *Alien Encounters: Anatomy of Science Fiction*. Harvard University Press, Cambridge, MA.

18. Wordsworth, W. (1798) in the Preface to *Lyrical Ballads*, quoted in Mark Rose's *Alien Encounters: Anatomy of Science Fiction* (1982). Harvard University Press, Cambridge, MA.

19. Rose, M. (1982) *Alien Encounters: Anatomy of Science Fiction*. Harvard University Press, Cambridge, MA.

20. Ferris, T. (1988) *Coming of Age in the Milky Way*. William Morrow, London.

21. *Ibid*.

22. Carey, J. (1995) *The Faber Book of Science*. Faber & Faber, London.

23. 'Avertissement de l'editeur', prefixed to the 1866 edition of *Voyages et aventures du capitaine Hatteras*.

24. James, E. (1994) *Science Fiction in the Twentieth Century*. Oxford University Press, Oxford.

25. Alkon, P. (1994) *Science Fiction Before 1900: Imagination Discovers Technology*. Twayne, New York.

26. Rose, M. (1982) *Alien Encounters: Anatomy of Science Fiction*. Harvard University Press, Cambridge, MA.

27. Verne, J. (1864) *Journey to the Centre of the Earth* (transl. F. A. Malleson, 1877). London.

28. Ferris, T. (1988) *Coming of Age in the Milky Way*. William Morrow, London.

29. Toulmin, S. (1982) *The Discovery of Time*. Chicago University Press, Chicago.

30. Rose, M. (1982) *Alien Encounters: Anatomy of Science Fiction*. Harvard University Press, Cambridge, MA.

31. Verne, J. (1864) *Journey to the Centre of the Earth* (transl. F. A. Malleson, 1877). London.

32. *Ibid.*

33. *Ibid.*

34. Aldiss, B. (1986) *Trillion Year Spree*. Victor Gollancz, London.

35. *Ibid.*

36. *Ibid.*

37. Price, L. (1956) *Dialogues of Alfred North Whitehead*. Signet, New York.

38. Carey, J. (1995) *The Faber Book of Science*. Faber & Faber, London.

39. Bernal, J. D. (1965) *Science in History*, Vol. II. C. A. Watts & Co., London.

40. Tennyson, A. (1885) Vastness, quoted in Mark Rose's *Alien Encounters: Anatomy of Science Fiction* (1982). Harvard University Press, Cambridge, MA.

41. Hobsbawm, E. (1968) *Industry and Empire*. Penguin, London.

42. Frye, N. (1966) Varieties of literary utopias. In: *Utopias and Utopian Thought* (ed. Frank E. Manuel). Houghton Mifflin, Boston.

43. Bulwer-Lytton, E. (1871) The coming race. In *Bulwer's Novels and Romances* (1897). New York.

44. *Ibid.*

45. *Ibid.*

46. Christensen, J. M. (1978) New Atlantis revisited: science and the Victorian tale of the future. *Science Fiction Studies*, #8, **5**(3).

47. Bulwer-Lytton, E. (1871) The coming race. In *Bulwer's Novels and Romances* (1897). New York.

48. Butler, S. (1872) *Erewhon*. London.

49. *Ibid.*

50. Butler, S. (1912) *The Note Books of Samuel Butler* (ed. H. F. Jones). Jonathan Cape, London.

51. *Ibid.*

52. Carey, J. (1995) *The Faber Book of Science*. Faber & Faber, London.
53. *Ibid.*
54. Isaacs, L. (1977) *Darwin to Double Helix: The Biological Theme in Science Fiction*. Butterworth, London.
55. Wells, H. G. (1933) *The Scientific Romances of H. G. Wells*. Gollancz, London.
56. Wells, H. G. (1895) *The Time Machine*. Heinemann, London.
57. *Ibid.*
58. *Ibid.*
59. *Ibid.*
60. *Ibid.*
61. Wells, H. G. (1893) On extinction. In: *Chamber's Journal*, London.
62. Wells, H. G. (1895) *The Time Machine*. Heinemann, London.
63. Baxter, S. (1995) Further visions: sequels to *The Time Machine*. *Foundation*, **65**, 41–50.
64. Huntington, J. (1995) *The Time Machine* and Wells' social trajectory. *Foundation*, **65**, 6–15.
65. Wells, H. G. (1898) *The War of the Worlds*. Heinemann, London.
66. *Ibid.*
67. *Ibid.*
68. *Ibid.*
69. *Ibid.*
70. *Ibid.*
71. Rose, M. (1982) *Alien Encounters: Anatomy of Science Fiction*. Harvard University Press, Cambridge, MA.
72. Wells, H. G. (1898) *The War of the Worlds*. Heinemann, London.
73. Philmus, R. M. (1970) *Into the Unknown*, Los Angeles.
74. Wells, H. G. (1898) *The War of the Worlds*. Heinemann, London.
75. Wells, H. G. (1934) *Experiment in Autobiography*. Macmillan, New York.
76. Smith, D. (1998) *Correspondence of H. G. Wells*. Pickering & Chatto, London.
77. Huntington, J. (1999) My Martians: Wells' success. *Foundation*, **77**, 25–34.
78. Huntington, J. (1979) The science fiction of H. G. Wells. In: *Science Fiction: A Critical Guide*. Longman, London.
79. Rose, M. (1982) *Alien Encounters: Anatomy of Science Fiction*. Harvard University Press, Cambridge, MA.
80. Wells, H. G. (1898) *The War of the Worlds*. Heinemann, London.
81. Rose, M. (1982) *Alien Encounters: Anatomy of Science Fiction*. Harvard University Press, Cambridge, MA.
82. Aldiss, B. (1986) *Trillion Year Spree*. Victor Gollancz, London.

83. Dick, S. J. (1996) *The Biological Universe*. Cambridge University Press, Cambridge.

84. Guthke, K. S. (1990) *The Last Frontier: Imagining Other Worlds, from the Copernican Revolution to Modern Science Fiction*. Cornell University Press, Ithaca, NY.

85. Crowe, M. J. (1986) *The Extraterrestrial Life Debate, 1750–1900*. Cambridge University Press, New York.

86. Flammarion, C. (1872) *Lumen* (transl. B. Stableford, 2002). Connecticut.

87. Crowe, M. J. (1986) *The Extraterrestrial Life Debate, 1750–1900*. Cambridge University Press, New York.

88. Flammarion, C. (1872) *Lumen* (transl. B. Stableford, 2002). Connecticut.

89. Dick, S. J. (1996) *The Biological Universe*. Cambridge University Press, Cambridge.

90. Flammarion, C. (1872) *Lumen* (transl. B. Stableford, 2002). Connecticut.

91. Particularly Humphry Davy's *Consolations in Travel; or, The last Days of a Philosopher* (1830) and Bernard le Bovier de Fontenelle's *Entretiens sur la Pluralité des Mondes* (*Conversations on the Plurality of Worlds*, 1686).

92. Flammarion, C. (1872) *Lumen* (transl. B. Stableford, 2002). Connecticut.

93. *Ibid.*

94. *Ibid.*

95. Wells, H. G. (1898) *The War of the Worlds*. Heinemann, London.

96. Flammarion, C. (1872) *Lumen* (transl. B. Stableford, 2002). Connecticut.

97. Russell, W. M. S. (1995) Time before and after *The Time Machine*. *Foundation*, **65**, 41–9.

98. *Ibid.*

99. Bergonzi, B. (1975) in *H. G. Wells: Early Writings in Science and Science Fiction* (eds. R. M. Philmus and D. Y. Hughes). University of California Press, Berkeley, CA.

100. Wells, H. G. (1895) *The Time Machine*. Heinemann, London.

101. Flammarion, C. (1872) *Lumen* (transl. B. Stableford, 2002). Connecticut.

Chapter 3

PULP FICTION: THE
ASTOUNDING AGE

Although ending up resident in the USA for many years and becoming fully bilingual, twentieth century scientific icon Albert Einstein claimed never to be able to write adequately in English because of 'the treacherous spelling'. He never lost his distinctive German accent, summed up by his catch-phrase 'I vill a little t'ink', and H. G. Wells' following description of Mr Polly could have been written with Einstein in mind: '[with] little or no mastery of the mysterious pronunciation of English, and no confidence in himself... He avoided every recognized phrase in the language, and mispronounced everything'[1].

Modern folklore may assert that Einstein reinvented time, but the truth is, as we have learnt, H. G. Wells got there first. It is from this premise, that astounding developments in both science and its fiction drove each other forward, that we venture into the twentieth century. Wells' negative assertion that 'science hangs like a gathering fog in a valley, a fog which begins nowhere and goes nowhere, an incidental, unmeaning inconvenience to passers-by'[2] is not to be endorsed. Neither the role that science played in driving science fiction forward nor the role that science fiction played in driving forward science is to be underestimated in this generation.

Einstein's Mirror
The physics of the early twentieth century feeds back into fiction. Newton's clockwork Universe is shattered by Einstein's vision of an expanding Universe. The glowing future suggested by the

rockets flying across the covers of the pulp magazine *Astounding Science Fiction* owe their new-found frontier to Hubble's deep space and a space–time stripped of temporal parochialism. Such is the synergy of science fiction and science. Science Fiction began this era with bug-eyed monsters, exploding galaxies and stories written like engineering diagrams. However, as the Astounding Age proceeded, its excesses were stripped away. Its science became coherent, and style and flair are to be found in the inky depths of the printed page.

Einstein redressed the stage upon which science fiction was played out. Prior to Einstein's new broom, Galileo's principle of relativity was dominant – a worldview which had inverted the absolutist opinions of Aristotle. Galileo asserted that motion in a straight line only had meaning relative to something else. Consequently there was no absolute reference frame by which all things could be measured. Galileo's model five laws of motion were slashed to three by Newton whilst still concurring on the existence of that same absolute reference frame. Reserving the principle of relativity for observed objects, uniform motion could not detect Newton's absolute space.

It was the Scottish mathematical physicist James Clerk Maxwell who developed a set of equations expressing the basic laws of electricity and magnetism, correlating the fact that changes in magnetic fields could transform electric fields and vice versa, in such a way that a solution for propagating electromagnetic waves could be set up. Maxwell's special role in the development of Einstein's Theory of Special Relativity led Einstein to describe Maxwell's work as the 'most profound and the most fruitful that physics has experienced since the time of Newton'[3]. Einstein incorporated the electromagnetism of Maxwell into the 1905 'Special Theory Of Relativity' outlined in *On the Electrodynamics of Moving Bodies*. Exploding Newtonian notions of space and time, Einstein re-weaves them together with the speed of light, creating his vision of special relativity: that time can pass more slowly if an observer is moving, depending on the relative speed of the observer and object. It was this theory that also produced the world's most famous equation, $E = mc^2$. It was aged 16, that Einstein set in motion this theoretical process when, after glimpsing his reflection in a mirror, he wondered what would happen to his simulacrum if he were rocketing forward at the speed of light; his conclusion – the speed of light is independent of the observer. His 'what if' effectively created a science fiction scenario.

The Twin Paradox – Paul Langevin

Science fiction was being created from Einstein's theory and Einstein wasn't the only one doing it. The French Physicist Paul Langevin, famed both for his work in echo location and for being the married lover of Marie Curie, is an unlikely creator of science fiction. Yet it was he who in 1911 created the Twin Paradox. This conceptual exercise has identical day-old twin brothers cruelly separated at birth as one is hurtled across the Universe by rocket whilst the other waits abandoned on Earth. When the cosmic voyager returns he is younger than the brother who remained behind. Langevin's explanation was that the twin who travelled was subject to acceleration whilst the forsaken brother was not. The French paramour's comment was that this is a paradox because the acceleration involved is absolute.

Einstein demonstrated that Langevin's logic was erroneous in his 1915 publication on General Relativity. Thus the only paradoxical element that remains of the Langevin story is that despite loving and leaving Marie Curie, evoking the wrath of a xenophobic French populace on Curie's head, her granddaughter still felt able to marry into the family by becoming wed to Langevin's grandson.

Einstein's theory of General Relativity curves both space and time to make gravity a feature of the Universe and not a force within it. Max Born, Einstein's assistant at the time he formulated his theory, comments on the fiftieth anniversary of Einstein's 1905 golden year: 'The theory appeared to me then, and still does, the greatest feat of human thinking about nature,... but its connections with experience were slender. It appealed to me like a great work of art, to be enjoyed and admired from a distance.'[4].

Born's comments are apposite, for there appears to have been a wider cultural attempt to grapple with the questions Einstein raised within this period of history. Artists were beguiled by the idea of the fourth dimension and transfixed by musings of what lay beyond the tain of Einstein's mirror. The world's creative minds grappled with how to represent such innovation. Visually there is a concurrent explosion in modern art with the development of cubism. The relativity of Einstein, with its exploration of the nature of space, is complimented by Picasso's explorations of the relation between perception and reality[5]. Picasso's 1907 *Les Demoiselles d'Avignon* explore the fourth dimension and offer insights into the parts of the Universe not directly observable to us. The relativity of Einstein, with its exploration of the nature of space, is complemented by Picasso's explorations of the relationship between perception and reality.

The scale of the Universe

The science fiction of this period takes Einstein's curved space–time as the sandbox in which to unleash its most incredible of concepts. Not content with opening up the Universe to a dazzling array of interstellar spacecraft, Einstein further strips the constraints from the authors of the Astounding Age and allows them free rein – not just of all space, but all time as well. In this, science fiction can explore the time dilation incumbent in the Twin Paradox, integrating the heartfelt anguish of the separated brethren and telling stories of a distant space traveller, caught at the perimeter of a interstellar black hole, who views time passing slower for the wristwatch tumbling into its depths when compared to the one upon his wrist, all based on the dilation found within General Relativity. At once the Universe seemed so large that any story possibility could exist somewhere, some time.

Just as the Universe was expanding for science fiction, so it was unfolding for science. On 26 April 1920, in the Baird auditorium of the Smithsonian Museum of Natural History, two influential scientists presented independent papers about 'The Scale of the Universe'. In a debate that was to soar beyond the conference and into the technical journals of the field, those two scientists, Harlow Shapley and Heber Curtis contended, with the true scale of the Universe as their prize.

Shapley's research pointed to the Milky Way as embracing the Universe in its entirety. Employing Dutch-American astronomer Adriaan Van Maanen's contention that because he had observed Andromeda rotating, its relative distance had been overestimated, it claimed that the speed of light was not constant. Curtis, in opposition, contended that Andromeda and her sister nebulae were 'island universes', separate galaxies from our own floating amidst the black void of the Universe.

Van Maanen's inaccurate observations, placing him in contradiction to Edwin Hubble's discovery that the Andromeda Nebula was a distant stellar object, were in error. Hubble proved that you cannot see Andromeda rotate during the brief span of a human life. Curtis's contribution – that the Milky Way is only one of hundreds of billions of galaxies hurtling away from each other in the visible Universe – was correct.

It was Hubble who expanded upon the work of the Roman Catholic Priest and astronomer Georges Lemaître. Based on Einstein's research, this Belgian cleric postulated that the Universe began with the explosion of a primeval atom. Hubble proved that on this basis the Universe was expanding. However, Einstein didn't believe in an expanding Universe, and had added a cosmological constant to his work to create an infinite and static Universe. He had to revise his thesis and drop the cosmological constant in what he himself described as his 'biggest blunder'. Perhaps this is why he was initially critical of Lemaître's work on what later came to be known as the 'big bang theory'. Yet by January 1933, Einstein, listening to Lemaître detail his hypothesis at a conference in California stood up at the end, led the applause, and remarked to those gathered there, 'This is the most beautiful and satisfactory explanation of creation to which I have ever listened'.

Dreams and legends – Soviet science and its fiction

Whilst some thinkers contemplated vast cosmic distances, others set their sights a bit closer to home. Whilst some were pondering the origins of the Universe, others were concerned with how we would travel to the nearest of our cosmic neighbours, the Moon. It is on the far side of that stellar body that the 75 kilometre Rynin crater sits. Perpetually bathed in twilight, like its namesake it is largely ignored by those who gaze upwards and dream of slipping the Earth's bounds. This magnificent site is named for Nikolai Alexsevitch Rynin, a Soviet engineer, teacher, aerospace researcher, author, historian, and relentless promoter of space travel.

This key figure in the history of science and science fiction cuts across these two fields and allows them to bleed into each other. He published a nine-volume encyclopaedia of space travel, *Mezhplanetnye Soobschniya* (*Interplanetary Flight and Communications*), between 1927 and 1932. This not only chronicled the research available internationally at the time, but also notes extensively the influence that science fiction played on the way that space travel was conceived. Rynin documents how the pulp fiction of dizzying technologies, immense interstellar starships, and fleet and nimble faster-than-light vessels worming through space–time, proved seminally influential on the space race in science fiction's Astounding Age.

A 'relentless propagandist'[6], Rynin was part of the Soviet passion for space travel. A passion which held 'a genuine appeal for first-generation Bolshevik technocrats who defined themselves as harbingers of the future and envisioned a utopian social order based on a foundation of advanced science and technology'[7]. This passion is well documented in the Bolshevik re-publication in 1919 of Alexander Bogdanov's 1908 science fiction novel *Red*

Star – an early space opera. In it a Russian Marxist travels to Mars to discover a socialist paradise – a paradise which illustrates the need for permanent revolution. For even after their plans had been realized the socialists of Mars realized that:

> During the most recent period of our history we have intensified the exploitation of the planet tenfold, our population is growing, and our needs are increasing even faster. The danger of exhausting our natural resources and energy has repeatedly confronted various branches of our industry[8].

Perpetually fighting a rising population, the socialist struggle in *Red Star* was set to continue. The theme of a socialist utopia on Mars is soon echoed in the worker-friendly mass medium of film. The classic 1924 Yakov Protazanov film *Aelita* portrays Soviet cosmonauts as the vanguard that leads the Martians in a successful revolution.

Although the first three of Rynin's volumes dealt with the dreams, legends and early fantasies[9] of spacecraft in science fiction[10], the majority of his work dealt with the theoretical development of space travel. Well read and well connected, Rynin demonstrates significant knowledge of developments outside the USSR, dealing with the German Hermann Oberth, the Frenchmen Robert Esnault-Pelterie, the father of Russian rocketry Konstantin Tsiolkovsky and the American Robert Goddard. Rynin even goes so far as to flatter Goddard by comparing his work favourably with his main European rival in a letter which states: 'I have read very attentively your remarkable book *A Method of Reaching Extreme Altitudes* edited in 1919 and I have found in it quite all the ideas which the German Professor H. Oberth published in 1924'[11].

Of cabbages and kings[12]
Robert Goddard stands as the epitome of the early American desire to conquer space. Despite public ridicule and rebuke

spanning decades, he could still be found, on a cold New England day, 16 March 1926, clutching the launching structure of his first liquid-fuelled rocket, named 'Nell'. Using his Aunt Effie's farm as a proving ground, Goddard launched Nell to see her rise just 41 feet in a 2½-second flight that ended in the vegicide of Aunt Effie's cabbage patch. Yet the fact that Nell broke free from the Earth's bounds for just a moment allowed the hopes and aspirations of a nation to also take flight, and with them, the developments which were to shape the Astounding Age. Goddard's point had been made – liquid-fuel propellants could be used to send a rocket surging skyward instead of exploding in a catastrophic detonation.

However, the motive force for Goddard himself was not drawn from the arcane glyphs and diagrams of science. Instead this luminary of space science claims that science fiction was his muse, writing in a fan letter to H. G. Wells:

In 1898, I read your *War of the Worlds*. I was sixteen years old, and the new viewpoints of scientific applications, as well as the compelling realism... made a deep impression. The spell was complete about a year afterward, and I decided that what might conservatively be called 'high altitude research,' was the most fascinating problem in existence[13].

To be so inspired by one of science fictions great authors was a common heritage that Goddard shared with Hermann Oberth the German pioneer. In Oberth's case interest was sparked at the age of 11 by Jules Verne's *From The Earth To The Moon*, a book that he asserted that he had read many times and even went so far as to say he knew it by heart. Despite Rynin's dismissive attitude to Oberth's contribution to rocketry, he shouldn't be underestimated. His innovative 1922 doctoral thesis detailing a liquid propellant rocket was initially rejected as being too unorthodox. Not dissuaded, Oberth published *Die Rakete zu den Planetenräumen*

('The Rocket into Interplanetary Space') on his own, a version of which won international plaudits and became not just a standard text, but also the inspiration for those who were to follow.

Not least amongst those inspired was Oberth's pupil, Wernher von Braun. A controversial figure by any measure, he was the main technical expert behind the Nazi 'retaliation weapon': the V2 rocket. Von Braun became naturalized in the United States following the Second World War, as part of their space program – all this despite controversy over his status as an SS officer and use of concentration camp prisoners as labour, a policy that saw more people killed by the production and delivery of the V2 than in their bombardment. Almost exactly 25 years after abandoning Germany, the Von Braun-designed Apollo 11 took off for the Moon propelled by the Saturn V booster. Rarely in the record of human enterprise has so stirring an aspiration been achieved by such ethically dubious means.

Proof that rocketry always produced casualties can be derived from the fact that Von Braun aided Oberth in his role as the technical advisor for the Fritz Lang science fiction film *Frau Im Mond* (Woman in the Moon). It was during an attempt to build a working rocket to drive publicity for the film that Oberth lost the sight in his left eye. Coupled with the assortment of injuries sustained by most of the early rocketry pioneers we can envisage a motley collection of revenants hoping to keep both body and mind together whilst lurching towards their final goal.

A *muse of fire*[14]

That both Oberth and Von Braun became involved in the cinematic industry was no surprise. Cinema was marked by the same level of innovation and drive that they themselves were used to in their rocketry circles. From its earliest beginnings science fiction films featured the application of practical science to create extrapolative yet fictional wonders. Early examples took as their core, ground breaking special effects which ignited those very first audiences. It was the French filmmaker/magician Georges

Figure 3.1 The iconic moonshot from Melies' 1902 film.

Melies who first extrapolated the techniques used within the field of illusion and magic then fused them with the emerging technology of film, basing his plotlines on both Jules Verne (*From the Earth to the Moon*) and H. G. Wells (*First Men in the Moon*). His pioneering 1902 work, the 14-minute *Le Voyage Dans La Lune* (A Trip to the Moon), literally *envisioned* a journey to the Moon by means of a craft fired from the bowels of a powerful gun. The space exploration storyline, coupled with its fantastic vision of a lunar surface populated by terrifying aliens and its pioneering special effects, influenced a whole genre of science fiction film in some way. Yet Melies cannot lay complete claim to being *the* science fiction film pioneer.

Otto Rippert's 1916 lost classic *Homunculus* introduced many characteristics repeated in later works. Not only was his work shown in sequential form over six episodes, predating the serial science fiction of the USA by some twenty years, but its plot of mad scientists, superhuman androids, dark shadows and sinister technology extrapolated the *Frankenstein* story onto film. It was *Homunculus*, rather than the earlier 1910 version of *Frankenstein*, that served as a more influential model for later film adaptations of Mary Shelley's iconic text. It also served as something of an introduction to the industry for one of the most influential figures in science fiction film. Rippert's screenwriter on *Homunculus* was Fritz Lang, the auteur who was to make *Metropolis*, one of the most expensive silent films ever completed.

Metropolis (1927) was the first significant science fiction feature after the First World War. Building on the horrific accounts of how gas, tanks and machine guns had decimated a generation of European young men, it demonstrated the potentially destructive effects of technology when it spins out of control. Its evil scientist *Rotwang* exerts his authoritarian control through the manipulation of oppressed enslaved underground industrial labour by the menacing yet magnificent metal Maria, a female robot facsimile of the pinup of the workers in the ultramodern city. Lang's work ran so over budget that it made huge losses, yet this did not deter him from pursuing a science fictional vision of the future with his next great work, released just two years later.

This 1929 science fiction epic, *Frau Im Mond* (Woman in the Moon), shows the first countdown to launch a rocket. Not just the first countdown on film, but the first countdown ever, anywhere. The concept of counting down as a coordinating feature was invented as a dramatic device for this film. *Frau Im Mond* also demonstrated for the first time on screen the use of liquid rocket fuel, two-stage rocketry, and zero gravity in space. Indeed, when the Nazis, under Wernher von Braun, began working on the military applications of rocketry, they decided the technology on display in *Frau Im Mond* was too close to their developments. In an effort to preserve the secret nature of their project, Hitler ordered the film's models be destroyed and had the film withdrawn from release.

Lang's works marked the evolution of the early science fiction films from a European perspective. Emphasizing prediction and social commentary, filmmakers like Lang sought to incinerate the veneer of society, exposing the truth beneath. By contrast, the American science fiction films of this period were all smoke and no fire at all, emphasizing daring exploits, melodrama and the application of technology. The American science fiction films of the 1920s would lead inexorably to the classic serials of the 1930s, *Buck Rogers* and *Flash Gordon*.

Such melodrama was the result of a nation seeking to escape from the realities of the Great Depression. The *talkies* replaced the manic over-the-top physical expressionism of the silent era with introverted features focusing on dialogue and character-heavy portrayals. Science fiction was no exception, partly because, despite Lang's pioneering work in the field, special effects were still primitive. The period's only exception to this trend was *King Kong* (1933), which featured a primeval primate plucking aeroplanes from midair whilst perched atop the symbol of both democracy and capitalism. When companies tried to imitate the vast scale of *Metropolis*, they failed at the box office, as *Metropolis* itself had done; this despite the fact that United Artists' Alexander Korda-helmed *Things to Come* (1936) was an adaptation of a 1933 H. G. Wells story, *The Shape of Things to Come*, which told of scientists trying desperately to save the globe with technology.

Instead, the public imagination was to find its focus in the rise of serial movies. These popular sequences were the low-budget, often comedic, interstellar romances shown as cliffhanger Saturday matinée fodder. Heroes like Dick Tracy, always battling mad scientists with his high-tech gadgets to foil their plots for world domination, and Tarzan, battling the poachers and renegades of the jungle, are joined by science fiction's first pair of iconic screen heroes, Flash Gordon and Buck Rogers.

The link between the two heroes lies with the choice of actor who was to portray them. Larry 'Buster' Crabbe was a former Olympic gold medallist swimmer whose blond hair and rugged good looks were utilized by the studios to bring these two intergalactic heroes to the big screen. Of the two, the more popular on screen was Flash Gordon. A rugged hero whose battles against Ming, Emperor of Mongo were aided by Flash's sweetheart Dale Arden and the scientific brains to Flash's brawn – Dr Hans Zarkov. Based on the 1934 Alex Raymond comic strip of the same name, its cinematic success repeated the popularity of the comic strips from which it was drawn. Flash was written as a response to the more popular *Buck Rogers in the 25th Century*, by Dick Calkins, a character which first appeared in the 1928 novella *Armageddon 2419 AD* from *Amazing Stories* magazine. It was on 7 January 1929, when the comic strip adaptation debuted, that Buck Rogers came to be a runaway success. Rogers, left in suspended animation through an accident, emerges in the twenty-fifth century. There, alongside his new friends Wilma Deering and Dr Huer, he battles both evil warlords and hordes of rocketship-wielding barbarians.

Vast fleets of rocketships marked the arrival of both Killer Kane, Buck Rogers' nemesis, and Ming of Mongo, the antagonist of the *Flash Gordon* serials. These curvaceous vessels cut through the sky bringing death and destruction with them. Seen as the ultimate military application, they were the source of the authoritarian power of the despot, a situation that was not too far removed from the reality of rocket research itself. In this, science fiction directly reflects the growing concerns of science with the potential power of rocketry.

Rockets' red glare[15]

In 1928 Hermann Oberth received the first Prix REP-Hirsch, a newly instituted annual prize of 5000 francs for the most outstanding work on astronautics. Although awarded by the Astronautical Committee of the Société Astronomique de France (the organization founded by Camille Flammarion), in truth it was the brainchild of the banker André Louis-Hirsch and the French rocket pioneer Robert Esnault-Pelterie. Esnault-Pelterie was so impressed by Oberth's work and optimism that this first prize was doubled to 10,000 francs. Esnault-Pelterie spent significant portions of his own life engaged in military work. This experience led him to the conclusion that, despite an obviously fertile imagination, one could not conceive of any defence against a strategic missile deployment. He commented that based upon his extensive experience of rocketry they could deliver 'over several hundreds of kilometres... thousands of tons of destructive payload, all within a few hours'[16]. Moreover 'the necessary ground installations would not entail great expense and would doubtless be infinitely less burdensome than if it were a question of delivering the same load by aeroplanes'[17]. Ming the Merciless would have been proud.

Esnault-Pelterie's association with space travel began in early 1912, when he lectured on its possibilities in both St Petersburg and Paris, the latter to the Société Française de Physique (SFP). The SFP was the most prestigious organization to provide a forum for what at that time many people still considered a fantastic subject. Esnault-Pelterie was, through his speculation, creating a fiction of the science, albeit couched in the terms of a public talk. He was creating an early piece of science fiction. Such was the visceral response to his comments that 'only Esnault-Pelterie's reputation in other fields permitted him to lecture'[18]. His later presentation of 1927 to the SFP was greeted with more respect despite being even more audacious, a sure sign that the science was catching up with the fiction. Indeed, the SFP subsequently published the written text as a 98-page book, *L'exploration par fusées de la tres haute atmospher et la possibilité des voyages interplanetaires* ('Rocket Explosion of the Very High Atmosphere and the Possibility of Interplanetary Travel').

This optimistic title, more daring than the 1912 lecture, combines Goddard's reticence and Oberth's confidence – an optimism which wasn't even dented when a rocket explosion left him without fingers on his left hand. Despite the disability his creativity led to the successful filing of over 120 patents, including one for the joystick. It is this sense of creativity that made him so successful.

Last and first men – Konstantin Tsiolkovsky

An apocryphal story tells of how Esnault-Pelterie in his 1905 trip to St Petersberg debated the future of space exploration with the

Russian rocket pioneer Konstantin Tsiolkovsky in the presence of the Tsar[19]. Although both participants of this debate denied the story, it is possible that Tsiolkovsky was aware of Esnault-Pelterie through a visit to Paris at the end of the nineteenth century to gaze upon the Eiffel Tower. Staring intently at this magnificent structure thrusting into the Paris skyline he conceived a method to make space travel inexpensive. This revelation was an elevator. A tower that would reach into space. Tsiolkovsky's vision was partly a reaction to the work of the English clergyman Thomas Malthus and partly the consequence of his attraction to the philosophy of Nikolai Fyodorovich Fyodorov (1829–1903). Fyodorov, a Russian Orthodox Christian philosopher, whom Tsiolkovsky had met whilst living in Moscow, was part of the Russian cosmism movement. He advocated radical life extension using scientific methods, human immortality and resurrection of dead people.

Fyodorov was a futurist who reflected on the eventual perfection of humanity and society leading to a Utopic existence. His writings included radical ideas on immortality, the revival of the dead, and both space and ocean colonization. He stands in contradistinction to Malthus who, in his *Essay on the Principle of Population*, published in 1798, made his infamous prediction that the world's population would outrun food supply. Malthus favoured moral restraint as a solution, but only for the working and poor classes:

> The power of population is so superior to the power of the earth to produce subsistence for man, that premature death must in some shape or other visit the human race. The vices of mankind are active and able ministers of depopulation. They are the precursors in the great army of destruction; and often finish the dreadful work themselves[20].

Malthus's views were conceived as a reaction to the optimistic views of his father and his associates, most notably William Godwin, father of Mary Shelley, the author of *Frankenstein*.

Tsiolkovsky reasoned that with inexpensive and regular access to space and the other planets, it would be possible for humanity to extend its reach into space, thus avoiding the devastation predicted by Malthus so that:

> The finer part of mankind will, in all likelihood, never perish – they will migrate from sun to sun as they go out. And so there is no end to life, to intellect and the perfection of humanity. Its progress is everlasting[21].

Tsiolkovsky's most famous work is the first scholarly treatise on rocketry, *The Exploration of Cosmic Space by Means of Reaction Devices* (1903). Yet during his life he authored over five hundred works on space travel and related subjects. These include his science fiction novels *On The Moon* (1895), *Dreams of the Earth and Sky* (1895) and *Beyond the Earth* (1920). His works may seem at first glance to meander without focus across the whole field of rocketry. Nestling amidst the designs of rockets are plans for steering thrusters, multi-stage boosters, space stations, airlocks, and essential systems to provide for the successful operation of space colonies. Yet attempts to flee Malthusian devastation lie always at the core of his work.

Under the influence of the philosopher Fyodorov, Tsiolkovsky believed mankind would evolve into a starfaring species, a concept explored not just in science but in its fiction as well, and most particularly in immense detail in the work of the innovative and influential philosopher and British science fiction author Olaf Stapledon. His 1930 work *Last and First Men* is an anticipatory history of humanity, whilst his 1937 novel *Star Maker* is an outline history of the Universe. Both synthesize contemporary astronomy and evolutionary biology to form myths apposite to a sceptical and scientifically cultured Astounding Age. Gregory Benford comments in an introduction to *Last and First Men* that Stapledon took his own study of evolution and 'projected it onto the vast scale of our future, envisioning the progress of intelligence as another element in the natural scheme'[22].

The transhumans of Stapledon's novels cross voids between the stars in vessels unimaginable to moribund earth dwellers. Viewed from the confines of a pedestrian existence these vessels journeyed the skies above in both architecture inconceivable and technology incredible. Science fiction had invented the UFO. Witness Stapledon's account of the Martian invasion of Earth:

> early walkers noticed that the sky had an unaccountably greenish tinge, and that the climbing sun, though free from cloud, was wan. Observers were presently surprised to find the green concentrate itself into a thousand tiny cloudlets, with clear blue between... though there was much that was cloudlike in their form and motion, there was also something definite about them, both in their features and behaviour, which suggested life[23]

It was not until after the Second World War that science started to explore the consequences of physical contact with an alien intelligence. But science was playing catch-up with the potentialities already explored by science fiction authors like Stapledon. The public at large might have already become used to the idea that Earth was located in the margins of the Universe, but it was science fiction that was preparing the public for their final demotion from the centre of the Universe – a demotion found in first contact between the terrestrial and extraterrestrial.

Pulp fiction and science communication

It is in attempts to institute a physicality of contact that the science fiction of the early twentieth century dwells. Konstantin Tsiolkovsky's science fiction publishing career began with his first science fiction story, *On the Moon*, in an 1892 Moscow magazine[24]. This attempted to make manifest the transhuman aspirations found in his science writing. The unidentified other that was explored in Tsiolkovsky's story was a common theme throughout the world in the mass magazines that developed in the 1890s. Together, these new periodicals, the pulp fiction magazines of the turn of the century, published many stories of science, romance and wild adventure, stories such as Edgar Rice Burroughs' Mars series, Ellis Parker Butler's detective stories and the dime Westerns of Zane Grey. It is in the pages of these daring exploits that the unidentified is catalogued. Whether they are American Indians or dwellers from another planet, the effect is the same.

It is clear from Rynin's encyclopaedia that he was familiar both with these magazines and with the early pulps that were starting to develop in America. His first three volumes demonstrate his familiarity with much of the European and American science fiction of the period. In fact, they indicate that there was considerable international influence and interchange of ideas in the genre. Indeed, John J. Pierce comments that 'Gernsbackian SF seems to have been an international movement'[25]. Here he is referring to Hugo Gernsback (1884–1967), the American editor, publisher, inventor, author, and founder of the first science-fiction magazine, Amazing Stories, in 1926. Gernsback's work in the burgeoning Amazing Stories and prior to that in publications like Modern Electrics, was covered by Rynin along with the work of the Europeans Bruno H. Bürgel and Otto Willi Gail.

Gail (1896–1956), a German science journalist and writer, was well known for his association with the communication of ideas in early rocketry. In particular the work he contributed to mediating the overly technical ideas of Oberth into a form that the general public could understand. Such was the complexity of Oberth's work that it needed someone who was 'knowledgeable in the physical sciences and able to fathom the dense snarls of rocket mechanics for ordinary intellectuals'[26]. Gail was ideally suited, having studied physics in Munich. He worked extensively for newspapers and the broadcast media and contributed to books on physics, astronomy and space travel. Beside this, Gail wrote science fiction novels, works whose underlying ethic was to

Figure 3.2 Cover of the famous September 1928 *Amazing Stories*.

inspire the young people of the day to explore, including the 1925 *Der Schuß ins All* (The Shot Into Infinity) and the 1929 *Die Blaue Kugel* (The Blue Ball). Gail's The former was subsequently bought in translation by Hugo Gernsback and became the lead story of his new publication *Science Wonder Quarterly*, Volume 1, Number 1 in 1929[27].

It is because Gail was in constant contact with the space travel pioneers of the time, including Hermann Oberth, that his novels are characterized by huge amounts of technical knowledge and specific details. It is this desire to be accurate that leads to Gails appearance in Rynin's encyclopaedia. Rynin's attempt to communicate the evolution of rocketry is the ideal place to situate a science communicator and author like Gail.

Like Gail, Bruno H. Bürgel (1875–1948) was an author, scientist and science communicator. A noted astronomer, he chose to work at the Urania observatory in Vienna. This institution was constructed to the designs of Art Nouveau architect Max Fabiani at the mouth of the Wien River and opened by Emperor Franz Joseph as an educational centre with a public observatory. Named after the Muse *Urania*, who represents astronomy, it was dedicated to popularizing astronomy and communicating science in general. Burgel's role at the observatory included the practical transmission of science through lectures and demonstrations. It is perhaps through this role as a science educator that Bürgel rates an inclusion in Rynin's volumes. Burgel managed to pack in a mix of work through his life, ranging from his observational work to his practical science communication. He also wrote a popular 1908 astronomy handbook *Aus fernen Welten – Eine volkstümliche Himmelskunde* (*From Distant Worlds – A Popular Science of the Sky*). In this volume he gathered the extant astronomical knowledge and describes it in a populist way with a conversational tone. The many illustrations and style of writing were to influence his later creation of such science fiction stories, tales like *The Cosmic Cloud*, which appeared in Gernsback's *Science Wonder Quarterly* of 1931, only two years after Otto Willi Gail's novel had appeared. The pioneering work that was being created on the continent at this time was inspiring Gernsback to open his readers' eyes. Through these stories they saw not just the fantasy of space travel, but extrapolative harder edged science fiction driven by contemporary research and thinking.

Gernsback believed that fiction could be a means for the effective dissemination of science. Like Otto Willi Gail he saw it as a way to inspire the next generation of scientists. Gernsback's editorials make it clear that he published and wrote stories with this agenda. A classic example of this manifesto is Gernsback's own work *Ralph 124C41+*, first published in serial form in his popular science magazine *Modern Electrics* in 1911. In *Ralph 124C41+* ('One To Foresee For All'), Gernsback outlined developments later classified as radar, the direction finder, space travel, germicidal rays, microfilm, two-way television, night baseball, tape records, artificial silk and wool, stainless steel, magnesium as a structural material, and fluorescent lighting:

he was Ralph124C41+, one of the greatest living scientists and one of ten men on the whole planet earth permitted to use the Plus sign after his name. Stepping to the telephoto on the side of the wall he pressed a group of buttons and in a few minutes the faceplate of the Telephot became luminous, revealing the face of a clean shaven man about thirty[28].

Barnum of the Space Age – Hugo Gernsback

Hugo Gernsback is known primarily in science fiction circles as the 'father of science fiction'. He was officially accorded that title by the World Science Fiction Convention in 1960 when they named their annual award the 'Hugo' after him. To engineers, however, Gernsback was an early pioneer in radio and television broadcasting. His station, WRNY - 1010 AM, Coytesville, started broadcasting on 12 June 1925 from its New York studios at the Roosevelt Hotel at 45th Street and Madison Avenue. It was on WRNY that he began a regular sequence of 'radio television' experiments. In these Gernsback presented daily 5 minute programs via 48 line mechanical scanners broadcast through the WRNY transmitter. Over 2,000 viewers watched a schedule made up of cooking lessons, physical fitness instruction and concerts. As a communicator Gernsback was unparalleled in early science fiction, his obituary in the *New York Times* going so far as to remind its readers that he was referred to as the 'Barnum of the space age'[29].

On the surface, Gernsback seems to have led a scattered life. But all of his efforts focused on the future. Hugo Gernsback was the modern world's first futurist – one who not only speculated about the future, but also worked to make it happen and guide others to it. His defining contribution was as the publisher and creator of the pulp age of science fiction associated with the publication of mass market magazines. Gernsback was not only the progenitor of the way that we came to access this new genre, but also the originator of what we were to call it – coining the term 'scientifiction'[30], and subsequently developing it in 1929, with the founding of *Science Wonder Stories*, to 'science fiction'.

The early stories found in the pulps focused upon Gernsback's idea that such fiction could be used to inspire the next generation of scientists, the so called 'gadget SF'[31]. Yet the readership quickly became bored with the limited nature of stories set principally in the laboratory. They wanted excitement and adventure. This they gained with an evolution of the Edgar Rice Burroughs style of planetary romance, 'space opera', a fashion which came to dominate the pulp magazines for the period up to the Second World War. This fiction had been given the name by Bob Tucker in his 1941 fanzine *LeZombie*; his comment was:

> In these hectic days of phrase-coining, we offer one. Westerns are called 'horse operas,' the morning housewife tear-jerkers are called 'soap operas.' For the hacky, grinding, stinking, outworn space-ship yarn, or world-saving for that matter, we offer 'space opera.'[32]

Massive in scale, its square jawed heroes rescuing scantily clad and screaming heroines, it could at its worst reduce a reader to tears of frustration through bad plotting, superficial dialogue and hoary schemes. At its best it opened the vistas of the imagination to the possibilities of the Universe.

Ludism not Luddism – Cummings and Hamilton the world destroyers

The naïve optimism of Gernsback's gadget SF may seem laughable to us today. Yet in its time it is seen as one of the more serious attempts to extrapolate in fiction. Its ludic nature is present with us today precisely because of its incongruity with what we now know to be true. It was not written merely to entertain, unlike the morass of ludic tales that were to typify the prodigious amount of the output of the Astounding Age.

It seems at times that the pulps engaged in an over-reliance on characterization and the relational interplay between protagonists, an approach that relegates the science. It is important here to distinguish between space opera and planetary romance. Space opera is the logical inference of colonialist expansion by the western powers. It emerges from both the frontier western and sea adventure traditions. Both of these earlier forms focus on the inherent travel needed in attempts by dominant societies to impose a new culture upon all that they encounter, the former by wagon train, the latter by armada. In space opera it is the spaceship, the rocket, that becomes the tool by which mankind transgresses the Earth's atmosphere and the rights of the natives that he encounters. This contrasts with the planetary romance, which locates itself in the tradition of the lost world or lost civilization tradition and focuses less on the travel than the locale, normally an alien world. Less strident in its attempts to subjugate the local populace, the planetary romance tries to engender some characterization of the protagonists and their relationships instead of just reducing them to meaningless adversaries which need to be conquered.

One of the early founders of science fiction's new populist movement sits precariously balanced between the twin masts of planetary romance and space opera, never destined to play anything more than a supporting role. Ray Cummings began his association with the field early in a life which, although rife with hyperbole, did seem to incarnate some of the adventures about which he wrote. His background included living on orange plantations in Puerto Rico, striking oil in Wyoming, panning for gold in British Columbia, timber cruising in Alaska and acting as Thomas Edison's assistant for over five years[33]. He was a prolific author, with over 750 stories to his name (of which only about 150 are SF), including *The Girl in the Gold Atom*, *The Princess of the Atom*, *Tarrano*, *The Conqueror*, *The Man Who Mastered Time* and *Into the Fourth Dimension*. Despite being published in France, Spain and Japan, Cummings never received the recognition that other authors of space opera had, perhaps because, although initially innovative, Cummings failed to grow as an author. He may have introduced the use of naval terminology when it came to descriptions of spaceships and travel, but if you pick a Ray Cummings story at random from his life's work it is difficult to identify its year of origin. The relentless similarity of the texts starts to stupefy the reader over time. Indeed, a common theme of Cummings' stories, chemically induced human shrinkage, may seem innovative at first reading, but by the tenth time it seems to have been almost completely reduced in its ability to entertain. Science fiction writers of the pulp era were able to ascend their dizzying heights by climbing on Cummings' shoulders, whilst Cummings remained a pulp writer of space operas.

Cummings stands in comparison to Edmond 'world destroyer' Hamilton, who truly was a space opera author. A child prodigy, he went to study electrical engineering at college aged 14. However, his tender years meant that he left in his third year and took a job on the Pennsylvania Railroad whilst trying to figure out what to do with the rest of his life. Always interested in science fiction he commented that:

> My interest in s-f dates back to a time when I was so young I could barely read, but was fascinated by the illustrations of an H. G. Wells article, 'The Things that Live on Mars,' in the old Metropolitan magazine. I soon graduated to the Argosy and Allstory magazines, with the mighty Martian stories of Burroughs and all the other great fantastic stories they published for so many years[34].

Despite protestations that he had never shown any previous inclination towards writing, Hamilton decided in the mid-1920s to be an author. His first published short story, The Monster-God of Mamurth, was more of a fantasy tableau than a conventional SF tale. Yet as he developed he gradually found his feet in the space opera genre, his tales in which the hero defeats a major menace to the galaxy at the helm of a space armada through the destruction of a planet or two leading to his nickname.

Despite accusations that Hamilton's work suffers from a distinct lack of strategic plotting, he was also an innovator. He established a number of firsts in his works including the first use of a space suit in science fiction, the first space walk and the first use of an energy sword (the prototype for what George Lucas, a Hamilton fan, would later dub a lightsaber). He also found time to travel during this period, visiting much of the US and parts of Mexico in the company of his friend, author Jack Williamson. Williamson is another space opera author whose stories of this period are marked by a lack of concern for strategic plotting, the difference between Williamson and Hamilton being that the former had more time to hone his art, winning a Hugo award as a nonagenarian in 2001 having published his first story in 1928!

The superweapons found in the pages of a typical Williamson story and others like it were to have a remarkable effect upon the psyche of the American nation. As a consequence of the pulp fictions of Williamson and his contemporaries we are able to trace the emergence of a cult of the super weapon. The gleaming skyships of early space opera are the ultimate expression of mankind's destructive power. The American people, and in particular policy makers and risk takers like General Billy Mitchell, the decorated war hero and deputy director of the American Air Service, are affected by what they see in the fiction of their time. The fiction takes physical shape as developments in air power, from balloons to planes to missiles mean that 'America has led the way in transmuting the skies above us into the deadliest medium... inextricably intertwined with the development of these American super weapons have been those fantasies and illusions essential to their conception'[35].

Science fiction played multiple roles in legitimizing notions of air warfare. It was the fictional conception of air power as a super weapon which stimulated American culture and its people to accept the notion of air warfare before its practical creation. Science fiction created a vision of the superiority of air warfare over other forms (particularly naval). It legitimized the concept of aerial bombardment and created a code of acceptable war conduct which led to an emphasis upon strategic bombing. Space opera made it plain that he who rules the skies rules the land. Science was going to have to find a way to make the science fictional visions of the Astounding Age manifest in the wingspans of the aeroplane.

Doctors and donuts – E. E. 'Doc' Smith

Perhaps the best known and most accomplished of the pioneers of space opera is E. E. 'Doc' Smith. This food chemist not only invented the way to stick sugar to donuts, but also created some of the best known and written space operas of the pre-war years. His stories should be seen as the epitome of the space opera and feature relentless action sequences that transport the reader into the heart of interplanetary conflict. Unlike some of his contemporaries Smith's inclination was to extrapolate from existing technologies in his work, instead of embracing the completely illogical or impossible just for the sake of a story arc.

His *Lensman* sequence of novels, which began with the epoch covering *Triplanetary*, feature not just a plot spanning millennia, but also the first innovative use of non-terrestrial and non-human protagonists as heroes. Later mixed race and mixed species crews, like those of Gene Roddenberry's *Star Trek*, owe it all to Smith. Yet it was not just science fiction that owes him a debt.

Ideas from Smith's writings directly influenced American Military Strategy. Admiral Chester Nimitz, Commander in Chief of Pacific Forces for the United States and Allied forces during the Second World War, acknowledged Smith's influence, confirming that the idea of displaying all relevant tactical information through one coherent and integrated system was copied by the US Navy from the Lensman series' 'tank':

> Your entire set-up was taken specifically, directly, and consciously from the Directrix in your story. Here you reached the situation the Navy found itself in – more communication

Figure 3.3 A classic cover from *Science Wonder Stories* from 1939.

channels than integration techniques to handle them. In your writing you proposed precisely such an integrating technique and proved how advantageous it could be.[36]

Because of his typical reticence we only know this about Smith because his editor at *Astounding* magazine, John W. Campbell, shared the story around. Prior to becoming editor Campbell himself was a well known space opera author of the *Wade*, *Arcot* and *Morey* series of stories. However Campbell's most famous short story is the classic *Who Goes There?* In it he recounts the tale of Antarctic researchers who uncover a crashed alien ship, inside which is a vicious occupant. So impressive was the story that it was filmed not once but twice – first in 1951 as *The Thing from Another World* and then again by John Carpenter as *The Thing* in 1982.

It was as an editor that Campbell exerted the greatest influence. When asked by Isaac Asimov why he gave up being an author he remarked that by becoming an editor he has a creative hand in hundreds of stories instead of limiting himself to his own creations. His approach stressed scientific plausibility and he was many times reported as shouting, 'If you can't make 'em possible, make 'em logical. If you can't research it, extrapolate it!'. Campbell's prodigious imagination led him to commission pieces from both established and up-and-coming authors based around concepts of his own. It was the success of this approach that saw the definitive year of 1939 emerge as a key date within science fiction. Campbell's success at *Astounding* was spinning off many competitors (including *Science Fiction and Future Fiction*, *Startling Stories*, *Planet Stories*, *Comet*, *Stirring Science Stories*, *Dynamic Science Stories* and *Super Science Stories* all in 1939 itself[37]). He was also establishing a stable of emerging talent which was to both change the way science fiction was perceived and inspire entire generations of both scientists and scientific speculation. Campbell's *Astounding* introduced in 1939 *Black Destroyer* by A.E. van Vogt, *Life-Line*, Robert Heinlein's first story, *The Ether Breathers* by Theodore Sturgeon and *Trends*, the first effort by a teenage Isaac Asimov. Like Asimov, 1939 was the year that science fiction reached its maturity.

Science fiction was awkward and unsure of itself before, never quite comfortable with the mixture of ludic and extrapolative, lurid and exciting, which marked space opera. The advent of the Second World War and its far-reaching consequences on the entire planet marked the beginning of a consequent maturation of science fiction and ensured that both the science and style rocketed across the pages of the pulps, propelled by the sun-like bursts of atomic explosions.

References

1. Wells, H. G. (2005) *The History of Mr Polly*. Penguin, London.
2. Wells, H. G. (2005) *A Modern Utopia*. Penguin, London.

3. Einstein, A. (1931) Maxwell's influence on the development of the conception of physical reality, in *James Clerk Maxwell: A Commemorative Volume 1831–1931*. Cambridge University Press, Cambridge.

4. Max Born's comments on the 50th anniversary of Einstein's golden year.

5. Miller, A. I. (2001) *Einstein, Picasso: Space, Time, and the Beauty That Causes Havoc*. Basic Books, New York.

6. Golovanov, Y. A. (1994) *Korolev: Mify i Fakty*. Nauka Press, Moscow.

7. Crouch, T. D. (1999) *Aiming for the Stars: the Dreamers and Doers of the Space Age*. Smithsonian Press, Washington, p. 42.

8. Bogdanov, A. (2006) *Red Star*. Indiana University Press, Bloomington, IN, p. 79.

9. Rynin, N. A. (1970) *Interplanetary Flight and Communication Vol.1: Dreams, Legends, and Early Fantasies*. NASA and NSF, Washington DC.

10. Rynin, N. A. (1971) *Interplanetary Flight and Communication: Vol. 2: Spacecraft in Science Fiction*. NASA and NSF, Washington DC.

11. Burrows, W. E. (1998) *The New Ocean*. Random House, New York, p. 59.

12. Carroll, L. (1872) *Through the Looking-Glass and What Alice Found There*. Macmillan, London.

13. Goddard, E. C. and Pendray, G. E. (eds.) (1970) *The Papers of Robert H. Goddard*. McGraw-Hill, New York, p. 23.

14. Shakespeare, W. (1599) Prologue from *King Henry V*.

15. Key, F. S. (1814) The Star-Spangled Banner. In: *The Defence of Fort McHenry*.

16. Crouch, T. D. (1999) *Aiming for the Stars: the Dreamers and Doers of the Space Age*. Smithsonian Press, Washington, p. 36.

17. Von Braun, W. and Ordway, F. I. (1975) *History of Rocketry & Space Travel*, 3rd edn. Crowell Company, New York, p. 75.

18. *Ibid.*

19. *Ibid.*

20. Malthus, T. (1798) *Essay on the Principle of Population*. J. Johnson, London.

21. Tsiolkovsky, K. E. (1895) *Grezi o zemle i nebe* (Dreams of the Earth and Sky). Isd-vo, Moscow.

22. Benford, G. (1999) Preface to *Last and First Men*, by Olaf Stapledon. Millennium, London, p. x.

23. Stapledon, O. (1999) *Last and First Men*. Millennium, London, p. 131.

24. Griffiths, M. (2002) *Science Fiction – The Power Behind Spaceflight*. Unpublished paper presented at the British Rocketry Oral History Project, April 2002, Charterhouse Conference, Guildford.

25. Pierce, J. J. (1973) Rynin's Interplanetary Flight and Communication. *Science Fiction Studies*, **1**(2), Fall.

26. Burrows, W. E. (1998) *The New Ocean*. Random House, New York, p. 50.

27. Ashley, M. (2000) *The Time Machines*. Liverpool University Press, Liverpool, p. 65.

28. Gernsback, H. (1925) *Ralph 124C41+*. Cherry Tree, London, pp. 13–14.

29. 'Hugo Gernsback Is Dead at 83'. Obituary in the *New York Times*, Sunday 20 August 1967, p. 88.

30. Gernsback, H. (1926) Editorially speaking. *Amazing Stories*, 1 September 1926, p. 6.

31. Ashley, M. (2000) *The Time Machines*. Liverpool University Press, Liverpool, p. 231.

32. Tucker, W. (1941) *LeZombie*, January, No. 36, p. 9.

33. 'Raymond Cummings, Novelist, Dead'. Obituary in the *New York Times*, Sunday 24 January 1957, p. 29.

34. Hamilton, E. (1956) Autobiographical note. *Imagination Science Fiction*, April.

35. Franklin, H. B. (1988) *The Superweapon and the American Imagination*, Oxford University Press, Oxford, p. 8.

36. Smith, Trestrail V. (1979) *Keynote Speech At Moscon 1*, 29 September.

37. Carter, P. A. (1977) *The Creation of Tomorrow*. Columbia University Press, New York, p. 21.

Chapter 4

COLD WAR AND HEAT DEATH: THE ATOMIC AGE

In March 1944 a detail of counter-intelligence officers of the United States Army inspected the offices of *Astounding Science Fiction* magazine. Their brief was simple: to uncover potential leaks in the wake of publication of a fictional story about the development of an atomic weapon. The event entered science fiction mythology when the magazine's editor, John W. Campbell, later revealed his relief that the officers failed to detect his wall-map illustrating the distribution of subscribers across the US. Clearly marked on the map with bright-red pins was a cluster at PO Box 1663, Sante Fe, New Mexico – home to the Manhattan Project. Consternation at counter-intelligence HQ might have reached hysterical levels had they known that Wernher von Braun had imported a copy of the same publication into Germany for the duration of the war...

Atoms and void

The mysteries of the atom had long been a scientific Holy Grail. The Greek atomist Democritus was the first to suggest, around half a century before Aristotle, an eloquent case for the existence of elementary particles: 'Colour exists by convention, sweet by convention, bitter by convention; in reality nothing exists but atoms and the void'[1]. By the first light of the twentieth century it was clear that some form of atomic energy must be responsible for powering the Sun and the stars. In 1899 American geologist Thomas Chrowder Chamberlin was reasoning that atoms were 'seats of enormous energies' and that 'the extraordinary conditions which reside in the centre of the Sun may... set free a portion of this energy'[2].

The questionable credit for creating nuclear arms and foretelling nuclear Armageddon belonged not to science but to fiction. For the next four decades, nuclear weapons were to be found only in the pages of pulp magazines and mainstream science fiction. The fictional imagination has since been inextricably linked with the threat of real nuclear warfare. The best way to capture the Cold War climate is once more to look at the major players.

Foremost was the prolific H. G. Wells with *The World Set Free* (1914). In Wells' wake came the paranoiac fiction of the period. Leading this pack was the immensely influential *Nineteen Eighty-Four* (1948), Orwell's dystopian vision of life in the shadow of the Bomb. The mood of apocalyptic terror was tangible. George R. Stewart's *Earth Abides* (1949) impeccably portrayed the temper of the times. The prospect of annihilation and aftermath became the dominant theme in film and fiction, with Stanley Kramer's *On the Beach* (1959) and Walter M. Miller's *A Canticle for Leibowitz* (1959), reaching an apotheosis in Stanley Kubrick's *tour de force Dr Strangelove* (1964).

The world set free

In 1903 the great nuclear physicist Ernest Rutherford and his co-worker, Frederick Soddy, had been the first to calculate the vast amount of energy released in radioactive decay. Both were awake to the idea that this energy was potentially lethal. Indeed, Rutherford is alleged to have said, 'some fool in a laboratory might blow up the universe unawares'[3]. Soddy, in a lecture the following year, reasoned, 'The man who put his hand on the lever by which a parsimonious Nature regulates so jealously the output of this store of energy would possess a weapon by which he could destroy the Earth if he chose'[4]. Soddy trusted nature to 'guard her secret'. H. G. Wells begged to differ.

Wells' novel *The World Set Free* (1914) led non-stop to the launch of the Manhattan Project. The book features the building of what Wells here christens the 'atomic bomb':

> ... And these atomic bombs which science burst upon the world that night were strange even to the men who used them[5].

Ancient myths had professed that those of tainted spirit drinking from the Chalice would face instant annihilation. Wells was aware that the Holy Grail of the atom offered the opportunity for great good or sheer evil.

On the eve of the First World War, Wells presented an ill-omened vision of future warfare. The book foresaw a holocaust where the world's key cities are annihilated by small atomic bombs despatched from airplanes. This is no mere guesswork on the part of Wells. The weapons portrayed are truly nuclear: Einstein's equivalence of matter converted into fiery and explosive energy triggered by a chain reaction.

There had been earlier fiction on super-weapons. They had fallen prey to cliché; the naïve notion that the tangential mind of a single genius could change the course of history. Human problems could be solved by the techno-fix of a scientific miracle. Wells was wise enough to realize that the level of technical advance does not come from the know-nothing notion of genius. It comes from the dialectic between nations and their productive forces. Wells here predicted the emergence of the military-industrial complex.

His schedule for the development of nuclear capability is astoundingly accurate. In *The World Set Free*, the 1950s scientist who uncovers atomic energy realizes that there is no going back. Nonetheless, he feels, 'like an imbecile who has presented a box of loaded revolvers to a crèche'[6]. Initially nuclear capability merely leads to a greater strain on the system. The rich get richer, unemployment soars, crime rockets.

Global tensions become menacing, with governments 'spending every year vaster and vaster amounts of power and energy upon military preparations, and continually expanding the debt of industry to capital'[7]. These contradictions of capitalism, which Wells brands a barbaric form of society, lead to nuclear holocaust. The Earth is scorched. The swarms of survivors, many mutilated by fallout and radioactive dust, wander the barren landscape in scenes now familiar in film and fiction.

Science still emerges as 'the new king of the world'[8], however. In Wellsian fashion, a republic of mankind, governed by intellectuals and scientists, is established from the ruins of capitalism. Looking 'backward' from this post-apocalypse utopia to the mid-twentieth century, Wells delivers a damning indictment of his own time:

> They did not see it until the atomic bombs burst in their fumbling hands. Yet the broad facts must have glared upon any intelligent mind. All through the nineteenth and twentieth centuries the amount of energy that men were able to command was continually increasing. Applied to warfare that meant that the power to inflict a blow, the power to destroy, was continually increasing[9].

'I am become Death. The destroyer of worlds'

Wells' fictional bomb led straight to Hiroshima. His visionary novel was the guiding inspiration for the brilliant Hungarian physicist Leo Szilárd. After reading *The World Set Free* in 1932, Szilárd became the first scientist to seriously examine the science behind the creation of nuclear weapons. 'The book made a very great impression on me', Szilárd recalled. Thirty years later he still remembered the prophetic book:

> ... a world war... fought by an alliance of England, France, and... America, against Germany and Austria, the powers located in the central part of Europe. [Wells] places this war

in the year 1956, and in this war the major cities of the world are all destroyed by atomic bombs[10].

Szilárd was a survivor of a devastated Hungary after the Great War. He had developed a lasting humanitarian passion for the protection of life and freedom, particularly the freedom to communicate ideas. Wells' book echoed in Szilárd many of the utopian beliefs that guided him in the years to come. A quiet determination now changed the direction of his work. Szilárd went into nuclear physics because he wished to contribute something to save mankind: 'only through the liberation of atomic energy could we obtain the means which would enable man not only to leave the Earth, but to leave the solar system'[11].

A year after reading Wells' book Szilárd fled to London to escape Nazi persecution. There he read an article in *The Times* by Rutherford. The professed 'father' of nuclear physics, and pioneer of the orbital theory of the atom, Rutherford rejected the idea of using atomic energy for practical purposes. A legendary quick thinker, Szilárd was so incensed at Rutherford's dismissal that he dreamt up the idea of the nuclear chain reaction while waiting for traffic lights to change on Southampton Row in Bloomsbury, London. One year later he filed for a patent on the concept.

Szilárd became the driving force behind the Manhattan Project. It was his idea to send the confidential Einstein–Szilárd letter in August 1939 to Franklin D. Roosevelt outlining the possibility of nuclear weapons. The two brilliant and influential Jewish scientists feared the irresistible rise of a Nazi bomb. Within months the Manhattan Project was launched. It would ultimately boast over 130,000 employees, a total cost of $2 billion ($20 billion in today's figures), and the detonation of three nuclear weapons in 1945: the Trinity test detonation in July in New Mexico, a uranium bomb, 'Little Boy', detonated on 6 August over Hiroshima, and a plutonium bomb, 'Fat Man', discharged on 9 August over Nagasaki.

As the war raged on, Szilárd became sickened that scientists were losing power over their research. The military manoeuvres were sinister. He had hoped that the US government, resolutely opposed to the bombing of civilians prior to the war, would not use nuclear weapons on civilian populations. He hoped that the mere threat of such weapons would force Germany and Japan to surrender. So Szilárd led a petition, signed by 70 Chicago scientists, urging President Truman to demonstrate the bomb, not use it against cities as in *The World Set Free*.

Wells' nightmare became factual terror over Japan. As the 320,000 inhabitants of Hiroshima were waking up, the American Bomb airburst over the city. Within a second, thousands were slaughtered by the heat death, vaporized by the light and energy of the blast. Shadows on the walls were their only ghostly remains. They were the lucky ones. Victims further from the detonation were blinded, or had their skin and hair set ablaze. Later they would lose the white blood cells needed to fight the escalating disease.

Back in Los Alamos, many of the Manhattan Project scientists had celebrated news of the Hiroshima massacre. Austrian-British physicist Otto Frisch recalled how 'Somebody opened my door and shouted "Hiroshima has been destroyed"'. Frisch felt nothing but nausea when he saw how many of his, 'friends were rushing to celebrate. It seemed rather ghoulish', he thought, 'to celebrate the sudden death of a hundred thousand people'[12].

After their discovery had been used to blast two cities, Szilárd and the more humanitarian of the Manhattan physicists were left to doubt whether they should feel blessed or cursed[13]. Albert Einstein, a lifelong pacifist was, like Szilárd, a committed humanitarian. Einstein compared his fellow scientists to Alfred Nobel, the 'genius' behind explosives who was duty-bound to atone for his troubles. Einstein urged scientists everywhere to work for a world government, 'The war is won, but the peace is not'[14].

The Manhattan Project's lead scientist, Robert Oppenheimer, spoke for many physicists when he declared, 'In some sort of crude sense which no vulgarity, no humour, no overstatement can quite extinguish, the physicists have known sin; and this is a knowledge which they cannot lose'[15]. Oppenheimer believed that if atomic bombs were to be added as new weapons to the arsenals of a warring world, 'then the time will come when mankind will curse the names of Los Alamos and Hiroshima. The people of this world must unite or they will perish'[16].

The Cold War had begun.

George Orwell: champion of anachronisms

Szilárd's humanitarian atom bomb had burst in his fumbling hands. The relationships developed between physicists such as Oppenheimer, Teller, Bethe, Szilárd and Fermi laid the foundation for the speedy expansion of US science after the war. More than any other technical enterprise in history, the construction of the bomb illustrated one thing. Science, supercharged by huge resources, and infected by greed and the fear of rivals, as Wells foresaw, is capable of transforming the course of the world. Los Alamos marshalled a project with double potential. A weapon with which to win the war, and a weapon with which to win the

Figure 4.1 Big Brother watches over Winston Smith in the first cinematic rendition of George Orwell's *Nineteen Eighty-Four* from 1956.

peace in the post-war world. A world in which the politics of the twentieth century would be totally transformed.

George Orwell became the political spokesman of a generation that saw two world wars, economic slump, and incomparable suffering. Orwell's writing, culminating in *Nineteen Eighty-Four*, was greatly influenced by his relationship with two major authors of science fiction.

Aldous Huxley, creator of the hedonistic future *Brave New World*, was Orwell's English teacher at Eton. Together they would create contrasting classics of science fiction that defined the twentieth century debate on the question of science, power and politics.

When Orwell and Huxley first met in the fall of 1917 both these great books were as yet unwritten. While teacher and pupil roamed the tranquil halls of privilege, beyond England history was unravelling. Russia stood at the brink of Bolshevik revolution. Henry Ford, meanwhile, cranked up American capitalist production to turn out a million cars a year. The work of Orwell and Huxley would strive to mirror the fears and ideologies of this changing and contrasting world. Huxley feared that what we love would ruin us; Orwell feared that what we hate would ruin us. It was a discourse they would continue until their deaths[17].

Orwell had a lifelong fascination with science fiction. As a boy he had pored over the American pulp fiction of the Astounding Age, as well as the works of Jules Verne and H. G. Wells. As a writer he had a distinctly science fictional approach. In 1946 Orwell listed the things he stood for, '... socialism, industrialism, the theory of evolution... universal compulsory education, radio, aeroplanes...'[18]. The list points to Orwell's concern with progress and its social consequences. While he was an editor at the BBC, he showcased talks by practising scientists such as British evolutionary biologist J. B. S. Haldane and Irish-born physicist J. D. Bernal[19].

A post-war essay of Orwell's, *What is Science*, argued for a universal education in science and critical thinking. Throughout the

1940s he wrote on the cost of industrial advance and the huge contradiction present in the idea of industrialism:

> The tendency of mechanical progress is to make your environment safe and soft; and yet you are striving to keep yourself brave and hard. You are at the same moment furiously pressing forward and desperately holding back.... So in the last analysis the champion of progress is also the champion of anachronisms[20].

Orwell's other great influence was H. G. Wells. He noted the effect of Wellsian science fiction in inspiring a spirit of social change in his youth, 'back in the nineteen-hundreds it was a wonderful experience for a boy to discover H. G. Wells... here was this wonderful man who could tell you about the inhabitants of the planets and the bottom of the sea, and who knew that the future was not going to be what respectable people imagined'[21]. Indeed, the young Eric Blair adopted a name that alluded to Wells, admiring his work for its speculative brilliance and its ability to create 'a universe of its own'[22].

The right-minded Orwell was suspicious, however, of Wells' uncritical belief in science. In 1923 Wells had written the utopian novel *Men Like Gods*. The book describes a world in which science has led to the elimination of disease. Wells, like many intellectuals of his day, was a eugenicist. He felt that 'undesirables' should be treated as 'a malignant tumour; you would cut them out'[23]. To Wells his society of gods and perfect scientists was an ideal world. To Orwell it read like a fascist dystopia. In a BBC radio broadcast in 1942 Orwell characterized Wells as '... saying all the time, if only that small shopkeeper could get a scientific outlook, his troubles would be ended'[24]. Such blind faith in progress was limiting; Wells wore, 'the future round his neck like a millstone'[25].

Wells' most enduring political belief had been his idea of a World State. This society, such as in *The World Set Free*, would be a centrally planned meritocracy that would advance science. Nationalism would end, once and for all. For Wells there was no room for democracy. The average citizen could never be educated enough to resolve the major issues of the day. So the vote needed to be limited to scientists, planners, engineers, and, as can probably be guessed, Wells himself.

In his essay *Wells, Hitler and the World State*, Orwell was ruthless in his criticism of such ideas, 'Modern Germany is far more scientific than England, and far more barbarous. Much of what Wells has imagined and worked for is there in Nazi Germany'[26]. Damning enough. Orwell went on to condemn this approach as dangerously dogmatic, and the contents of Wells' book, *Guide to the New World*, as 'the usual rigmarole about a World State... federal world control of air power, it is the same gospel as he has been preaching almost without interruption for the past forty years'[27]. It was hardly surprising that Wells wrote to Orwell calling him a 'shit'[28].

Nineteen Eighty-Four

No novel written in the twentieth century has captured the popular imagination like *Nineteen Eighty-Four*. Orwell's haunting spectre is of big government gone mad with lust for power[29]. The very title of this classic dystopia became a cultural watchword. The word 'Orwellian' still ominously speaks of matters hostile to a free society.

No single work of science fiction has had a greater impact on politics. Big Brother is almost as famous as *Frankenstein*. He and the concepts of Room 101, Newspeak and the Thought Police are still with us in these days of euphemism and political spin. The novel is a mark of the revulsion with which Orwell viewed the bleak situation in post-war Britain. The mounting international stalemate caused by the bomb is portrayed to the extreme.

The key to Orwell's narrative is his belief in a 'catastrophic' future. It is a future of boundless despair. The book confronts the prospect of three totalitarian power blocks bringing history to a standstill. Big Brother is unassailable, 'If you want a picture of the future, imagine a boot stamping on a human face – forever'[30]. In his essay *You and the Atomic Bomb* published in October 1945, just months after Hiroshima and Nagasaki, Orwell had written with exceptional insight about the age of atomic weaponry that lay ahead. It was clear that he was preparing the heart of darkness within *Nineteen Eighty-Four*:

> We have before us the prospect of two or three monstrous super-States, each possessed of a weapon by which millions of people can be wiped out in a few seconds, dividing the world between them. It has been rather hastily assumed that that means bigger and bloodier wars, and perhaps an actual end to the machine civilization[31].

Orwell's vision was one of superpowers colluding and in tacit agreement never to use the bomb. In *Nineteen Eighty-Four* Eurasia is ruled by the Neo-Bolshevik Party; Eastasia by the Death-Worship Party; and Oceania, made up mainly of North America and 'Airstrip One' (the UK), is ruled by Big Brother, O'Brien and the Party of Ingsoc, or English Socialism.

Each superpower is evidently the archenemy of the other two. The truth is different. In fact the power blocs 'prop one another up like three sheaves of corn'[32]. Like the latter-day 'War on Terror', continuous and limited wars allow each super-state to maintain hysteria within their borders. By the time *Nineteen Eighty-Four* was in print the Cold War was a reality. Indeed, Orwell invented the phrase in his piece *You and the Atomic Bomb*. The Iron Curtain had fallen, and Orwell's account of the politics of power blocs was staggeringly perceptive.

The paradise of little fat men

The book became a mirror of the fears and frustrations of the individual caught up in a complex, overly rationalized society. It was a prophecy of a totalitarian future based on 'not any particular country, but the implied aims of industrial civilization'[33].

Orwell was troubled by the devastating effects of science and technology:

> Barring wars and unforeseen disasters, the future is envisaged as an ever more rapid march of mechanical progress; machines to save work, machines to save thought, machines to save pain, hygiene, efficiency, organization... until finally you land up in the by now familiar Wellsian *Utopia*, aptly caricatured by Huxley in *Brave New World*, the paradise of little fat men[34].

For Orwell, the advances in science described in *Brave New World* had neither sense nor purpose. There was no clear reason why society should be stratified in the complex way Huxley suggested. Science had made physical power unnecessary. Life had 'become so pointless that it is difficult to believe that such a society could endure'[35].

Huxley had imagined a technology that titillated. Orwell foresaw the technology of control. The insidious nature of *Nineteen Eighty-Four's* culture of surveillance stems from its telescreens and Thought Police. In Orwell's marvellous words, 'The Beehive State is upon us, the individual will be stamped out of existence; the future is with the holiday camp, the doodlebug and the secret police'[36].

The Party of Big Brother rationalizes language and perverts history. Time is tampered with, dates of events forgotten or unascertainable. The science of information is used to maintain political control, underlining Orwell's point that 'The really frightening thing about totalitarianism is not that it commits "atrocities" but that it attacks the concept of objective truth: it claims to control the past as well as the future'[37].

In *Nineteen Eighty-Four*, science's mastery of the machine is so complete that Utopia is possible. But poverty and inequality are maintained as a means of sadistic control. The visual medium of monitoring in the two-way telescreen is a brilliant evocation of the all-seeing eye. In Orwell's book it is politicized into a techno-logical nightmare. As Winston Smith dutifully follows the daily exercises on the telescreen, he is at the same time observed by it[38].

The system of technological surveillance in Orwell's fiction swiftly became fact. *Nineteen Eighty-Four* developed into the stan-dard text for describing the militarization of life. In 1954, US histo-rian of science Lewis Mumford declared the world of Big Brother to be 'already uncomfortably clear'[39]. American social scientist Wil-liam H. Whyte cited Orwell's influence in his 1956 *The Organiza-tion Man*, a best-selling study of corporate dictatorships such as General Electric and Ford. Sociologist David Riesman credited the popularity of dystopias to Orwell and the Bomb: 'When govern-ments have power to exterminate the globe, it is not surprising that anti-Utopian novels, like *Nineteen Eighty-Four*, are popular, while utopian political thought... nearly disappears'[40].

Novel super-weapon

Nineteen Eighty-Four is a flawed masterpiece. Orwell had identi-fied a new dark age. He saw the necessity for a social side to technological progress and to 'reinstate the belief in human brotherhood'[41]. The Cold War had, however, already created the need for an ideological super-weapon. Orwell's book was used with little regard to the author's intention. Many critics[42] suggest that this was made far easier by the book's unrelenting portrayal of defeat. Neither Winston Smith nor the 'proles' have a ghost of a chance against the mechanical horror. The novel reinforces passivity rather than undermines it.

As a result, Orwell's intended warning had become a 'piercing shriek announcing the advent of the Black Millennium, the Mil-lennium of damnation'[43]. This shriek, distorted by the mass media, had frightened millions. The novel's ambitious British television adaptation alone, broadcast on BBC Television in the winter of 1954, was watched by an audience of over nine million viewers. The production was hugely controversial. Questions were asked in Parliament. Many viewers complained about the supposed subversive nature and horrific content. Rather than advance understanding, Orwell's book had become a prominent item in the propaganda of Hate Week. Millions continued to see the conflict of East and West in terms of black and white.

Shaken by the anti-Communist cant conferred upon *Nineteen Eighty-Four* by US Republican newspapers, Orwell issued a disclaimer from his hospital bed, shortly before his death in 1949: '... the danger lies in the structure imposed on socialist and on liberal capitalist communities by the need to prepare for total war with the USSR, and by the new weapons, of which of course the atomic bomb is the most powerful'[44]. For Orwell there was also danger in the acceptance of a totalitarian outlook by intellectuals of all colours: 'The moral to be drawn from this dangerous nightmare situation is a simple one: don't let it happen. It depends on you'[45].

Men come and go...

In the atomic age, the world lived with a mood of apocalyptic terror. Millions of people responded anxiously to the catastrophic visions of science fiction, such as George Orwell's *Nineteen Eighty-Four*. The apocalyptic atomic and hydrogen monsters had not been let loose by Big Brother. They were down to the huge advance in physics. But the impact of the Bomb was clear enough. The future wasn't what it used to be. Gone were the monorails, the silver suits, and the generation starships of the Astounding Age. In their place were catastrophe and the ominous image of the thermonuclear mushroom.

Science fiction had long imagined doomsday. The strange settings for the end of the world had begun with Mary Shelley. Human creativity in science morphed into unknown power, capable of destroying our entire species. So it was with *Frankenstein*, 'the first great myth of the industrial age'[46]. Victor wrestles with alien forces that 'might make the very existence of the species of man a condition precarious and full of terror'[47].

Eight years after *Frankenstein*, Mary published *The Last Man*, the first novel to describe the extinction of humans. Her book ends at the close of the twenty-first century, when, after societal progress, war unleashes a worldwide plague[48]. The last survivor roams a desolate planet in a futile search for another living soul.

Likewise, in George R. Stewart's *Earth Abides* (1949), the catastrophe is a global disease with no cure. Stewart shuns the atomic obsession of the age; he wipes out most of humanity with a plague. Only those naturally resistant, or geographically isolated, survive. In this primitive apocalypse of man versus nature, Stewart is ambivalent as to the cause. The devastation may have been triggered by human error, an 'escape, possibly even a vindictive release, from some laboratory of bacteriological warfare'[49]. Or perhaps it is a natural process of decay, caused by the 'biological law of flux and reflux'[50].

Earth Abides, winner of the inaugural International Fantasy Award in 1951, overlaps with the anthropological history of California. In 1911, a gaunt alien figure of a man who spoke a strange tongue drifted out of the wilderness of the California mountains. He was 'discovered' by Dr Alfred L. Kroeber, celebrated Berkeley anthropologist and father of science fiction writer Ursula K. Le Guin. The mystery man was identified as the very last survivor of the European massacre of a Native American tribe, the Yahi of California. Though his real name was never known (because in his society it was taboo to say your own name), Ishi, meaning 'man' in the Yahi dialect, had his life famously chronicled in *Ishi in Two Worlds* (1960). He was the last free-living Indian in North America.

... but Earth abides

Stewart, who like Kroeber was based at Berkeley, chose as his main character for *Earth Abides* Isherwood, or Ish, who also emerges from the hills into an alien world. Overnight, modern society has been transformed into an agrarian post-apocalyptic landscape. Stewart provides great contrast to the brazen

Figure 4.2 Yahi translator Sam Batwai, Alfred L. Kroeber, and Ishi, photographed at Parnassus in 1911.

survivalist accounts that imagine civilization can be re-established in a few weeks with a semi-wrecked vehicle, a hacksaw and a Swiss army knife. *Earth Abides* suggests that civilization may never be the same again, the novel's title echoing the Book of Ecclesiastes: 'Men go and come, but Earth abides'.

The striking aspect of *Earth Abides* is its lack of post-apocalyptic clichés. There's no shortage of shelter or rations. Absent are the leprous biker gangs and the requisite roving mob of mercenaries. Neither is there a farcical final battle between good and evil. Instead, Ish is a scientist anti-hero; low on survival skills, high on brains. Stewart recognizes that even scientists are laden with human frailties, unlike the archetypal heroes of pulp fiction.

Earth Abides chronicles Ish's post-apocalyptic life. Ish's main effort is to restore society among the small band of plague survivors. The story focuses on the lasting positive effect on the ecosystem once the blight of industry has vanished. Indeed, at one stage, Ish sees a flickering *Coca-Cola* sign in the distance, and wonders how long the grid will keep alive this suspect symbol of civilization.

Just as the real-life Native American Ishi emerged from the wilderness as the last envoy of his tribe's culture, Ish is the last emissary of American civilization. The last American. *Earth Abides*, like many post-apocalyptic stories, appealed to commonly held secret desires. Sanctuary from the confines of civilized society, a less populated world, and the chance to test one's mettle against the elements.

Albert Einstein had said, in 1945, 'I do not believe that civilization will be wiped out in a war fought with the atomic bomb. Perhaps two-thirds of the people of the Earth might be killed. But enough men capable of thinking, and enough books, would be left to start again, and civilization could be restored'[51]. In contrast, Stewart's *Earth Abides* is an elegy to mankind. By the story's end, the community Ish founded has grown into a motley crew of superstitious hunter-gatherers, as primitive as the Neanderthal, and totally uninterested in rebuilding 'civilization'.

Post-apocalypse worlds

The post-apocalypse world became a staple fixture in science fiction. As heat death and its aftermath became a dominant theme, it also provided the setting for much major fiction of the day. *Lord of the Flies* is a case in point. Published in 1954, it was a reworking of an H. G. Wells' theme; civilization is only skin-deep. World War Three is a backdrop to the main story. The true horror, however, may not be due to an external threat, such as the Bomb, but innate human malevolence. If Golding was right, there was going to be trouble.

Admiral William Leahy, US Chief of Staff at the time of the 1945 detonations, suggested that the bombs were dropped to justify the $2 billion spent in their creation. He mortified many by claiming that the US, in being the first to go nuclear, 'had adopted the ethical standards common to barbarians in the dark ages'[52]. Leo Szilárd argued that a nation that sets such a precedent would have to bear the responsibility of heralding 'an era of devastation on an unimaginable scale'[53].

If it had happened once, it could happen again. Looking back, it is easy to mock the paranoia of the age. The US post-war obsession with communist spies was greatly influenced by the wartime espionage that gave the Soviets detailed intelligence on the Manhattan Project. German-born theoretical physicist and atomic spy Klaus Fuchs, among others, had provided Joseph Stalin with the blueprint to develop nuclear capability.

The first Soviet atomic bomb, exploded in 1949, was a direct copy of 'Fat Man'. English scientist C. P. Snow remarked that 'With the discovery of fission', inspired by Wells' fiction and triggered by Szilárd's chain reaction, '... physicists became, almost overnight, the most important military resource a nation state could call upon'[54].

The movie that saved the world

The Americans had believed themselves to be invincible in the aftermath of the war. So when Moscow entered the atomic age, Washington was stunned. As the 1950s wore on, the Cold War climate found its frosty reflection in film and fiction. There was much to report. McCarthyist anti-communist witch-hunts, the conspiratorial House of Un-American Activities committee, and growing international tension between superpowers equipped with the H-bomb.

A slew of popular science fiction B-movies captured the mood of the era. *The Day the Earth Stood Still* (1951) told the tale of an alien who visits Earth to warn global leaders that taking their quarrels into space will mean fatal costs. *Invaders from Mars* (1952), *War of the Worlds* (1953) and *It Came from Outer Space* (1953) cranked up the idea of imminent nuclear attack. *Invasion of the Body Snatchers* (1956) was a satire of McCarthyist paranoia in the early days of this ever-chillier Cold War.

Into this increasingly tense environment emerged the first film to take a serious look into the post-nuclear abyss. American director Stanley Kramer's *On the Beach* (1959) was adapted from the 1957 novel of the same name. The book was a post-apocalyptic plot written by British author Nevil Shute after he had emigrated to Australia. Indeed, the story is set in Australia, the only continent left unscathed by a nuclear war. Here too the fallout will soon engulf the land, and the story focuses on the various ways its key players face the looming devastation.

Shute had pictured a worse case post-nuclear scenario. No one escapes. As the survivors await the arrival of the radioactive dust, a few are tempted by the government's recommended cyanide capsules. A couple of wine connoisseurs lament the fact they have too little time left to drink their cellar's stock of port. One survivor takes part in a reckless car chase. When you've got nothing, you've got nothing to lose. Meantime, an American submarine re-surfaces in Melbourne. The sub had returned to its homeland to explore the source of a mystery radio signal coming from a devastated Stateside. Nothing is discovered. After a lone crew member was left ashore to roam the rubble of his hometown, the captain returned to Melbourne to revive a fleeting romance with a local girl.

Once these last traces of life have played out, the film ends as the camera pans the desolate suburbs of Melbourne. Finally, a poignant closing shot, a banner flutters in the breeze to a soundtrack of *Waltzing Matilda*. It reads, 'There is still time... brother'[55]. Alas, it seems the Australian legend that the film's female star, Ava Gardner, while shooting, glared around and said of Melbourne, 'What a great place to make a movie about the end of the world' is untrue[56].

On the Beach had a massive impact. The *New York Times* affirmed that, 'The great merit of this film... is the fact that it carries a passionate conviction that man is worth saving after all'[57]. American Nobel Prize winning quantum chemist Linus Pauling declared, 'It may be that in some years from now we can look back and say that *On The Beach* is the movie that saved the world'. Pauling himself went on to win the Nobel Peace Prize in 1962 for his crusade against nuclear testing. He remains one of the very few people to individually collect two Nobel Prizes.

The warmongers were furious. The *New York Daily News* was adamant that the film 'plays right up the alley of a) the Kremlin, and b) the Western defeatists and/or traitors who yelp for the scrapping of the H-bomb'[58]. *Time* magazine mocked, 'The picture actually manages to make the most dangerous conceivable situation in human history seem rather silly and science-fictional'[59]. Heaven forbid.

Shute's novel was based on the principle that fallout knows no boundaries, and that nuclear devastation will be complete. Kramer's film had tempered this fatalism. The movie's scientist, played rather bizarrely by Hollywood dancer Fred Astaire, launches a scathing attack on the belief in deterrence, 'Everybody had an atomic bomb and counter-bombs and counter-counter-bombs. The devices outgrew us. We couldn't control them'[60].

American nuclear physicist Edward Teller was fuming. As usual. He devoted an entire chapter of his book, *The Legacy of Hiroshima* (1962), to refuting Shute's novel. 'Although unrealistic', Teller reluctantly confessed that, 'Shute's elimination of any practical attempt to survive is frightening because it corresponds with the attitude of the overwhelming majority of our people'[61]. Indeed, the movie was so massively popular that Eisenhower's Cabinet discussed ways of replying to its message.

The US Government had considerable experience of propaganda campaigns. After the attack on Pearl Harbor in 1941, they had four years to develop a communication strategy. By the time of Hiroshima, ideas of fighting 'the enemies of freedom' and the need for military supremacy had been well established in the public psyche. At the start of the Cold War, US propaganda campaigns were simply a matter of replacing Nazism with Communism as 'the enemy of freedom'[62].

Mass media campaigns on the question of nuclear conflict made a message of survivability the dominant one. US propaganda contained only a positive message: practical measures to be taken in the event of a nuclear attack. The infamous *Duck and Cover* campaign was a case in point. Taught to generations of US schoolchildren from the late 1940s, once the trademark flash appears, you take cover – a table, or a wall perhaps – and simply cover your head with your hands. Hardly surprising that many thought the training to be of little help in the event of thermonuclear war. Indeed, such pro-nuclear messages merely increased the sense of paranoia.

Sputnik over America

Nevile Shute's *On The Beach* was the most famous of the 1950s anti-Bomb movies. It was greatly publicized, much debated, and very effective propaganda for the anti-nuclear lobby. But even its devotees had some doubts. In addition to the charges of defeatism and resignation, *On The Beach* was accused of 'sanitising war'[63]: 'The book and the film, by showing none of the physical agony and demolition that a real war could bring, made world extinction a romantic condition'[64].

The launch of Sputnik 1 in October 1957 sent shock waves across America. Future President Lyndon Johnson declared in a speech only months later that, 'Control of space means control of the world.... There is something more than the ultimate weapon. That is the ultimate position – the position of total control over Earth that lies somewhere in outer space...'[65]. The *Manchester Guardian* in the UK focused sharply on underlying American fears, 'The Russians can now build ballistic missiles capable of hitting any chosen target anywhere in the world'[66]. In future, apocalyptic fiction would hit far harder.

Helen Clarkson's *The Last Day* (1959) was a more considered account of nuclear war in the aftermath of Sputnik. The book tells of the trials of a middle-aged couple vacationing somewhere off the Massachusetts coast. The nuclear strike hits soon after

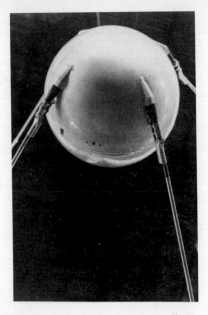

Figure 4.3 A model of Sputnik 1. The satellite's surprise launch shocked the United States.

their holiday begins. The book, like *On The Beach*, describes the creeping reach of deadly fallout on the wind. The novel's exposure of the sheer futility of civil defence measures won significant support in the US. The approval of Senator Clinton P. Anderson, Chairman of the Congress Joint Committee on Atomic Energy, adorned the book's cover. Having charged the AEC for years with secreting details on the dangers of fallout, Anderson praised the novel for injecting 'a little diet of realism' into the nuclear debate[67].

The Last Day tackles the aftermath of a nuclear raid. Through the voice of a half-dozen characters that eventually die, including the narrator herself, the story makes clear that there is no hiding place from the reach of fallout. In deliberate contrast to Shute's classic *On The Beach*, Clarkson presents a far less sanitized account of radiation death. The book details a medically accurate progression of symptoms, from sickness to complete loss of bodily functions.

The reader is continually presented with possible survival plots, only to have them demolished one by one. When a radio is repaired, fleeting hopes are raised of pending rescue:

> We would radio. Someone would send a plane. And some day some of those children would rebuild the world, a world where no one would ever make war again. They could never forget what had happened this time. There would be the mutants to remind them in each generation[68].

Once the radio is working, the reality is stark. Total radio silence sees such hope quickly evaporate. With delightfully catastrophic vision, Clarkson's chapter titles ('The First Day', 'The Second Day', and so on) count down the demise of the species as if 'Time itself was bleeding to death, second by second, and we could do nothing to staunch the flow from that mortal wound'[69]. The book dramatizes a bleak finale, and raises the question of ending, endemic to all nuclear war fiction, through the voice of one of the characters:

> I believe the most dangerous American tradition is the cult of the happy ending. We just can't believe that anything really bad can happen at the end of our story. We expect the going to be rough... but we have faith that everything will turn out all right in the end, no matter what we do'[70].

A Canticle for Leibowitz

Clarkson's book was a timely remedy to the chronic survivalist narratives of the decade. Pat Frank's *Alas, Babylon* (1959) was typical. Frank's book was notorious for depicting nuclear war as essentially winnable. *Alas, Babylon* is laden with the piety of US civil defence propaganda. The family will stay as society's nucleus and the house its refuge[71]. Frank's novel presents the post-apocalypse as severe but survivable. The government continues to run the areas uncontaminated by fallout, while helping survivors in 'contaminated zones'.

In the foreword of the 2005 edition, NASA consultant and physics professor David Brin confessed that the novel was decisive in forming his views on nuclear war. Indeed, the US Government had also been impressed with the power of Frank's propaganda. Civil Defence authorities used the book to guide local officials in ordering provisions in the event of a real nuclear attack. Dangerous talk.

The anti-nuclear campaign grew apace. In an article entitled 'Russia, the Atom and the West', published in the *New Statesman* on 2 November 1957, British writer J. B. Priestley had inspired the formation of the Campaign for Nuclear Disarmament. The first wave of unilateralism between 1958 and 1962 included many prominent founding members. Philosophers such as Bertrand Russell were joined by historians E. P. Thompson and A. J. P. Taylor, and renowned science fiction publisher, Victor Gollancz.

Science fiction had been instrumental in the development of the apocalyptic threat. Now it was time for critical insight. At this focal moment in history, many felt that science fiction was alone in its ability to project ways out of this predicament. It became the means by which a mass audience was confronted with the possibility of holocaust and mutually assured destruction. No other literature came close.

A Canticle for Leibowitz, published in 1959, is a post-apocalyptic classic and widely considered a masterpiece. It has never been out of print. *Canticle's* author, Walter M. Miller Jr, worked as an engineer during the Second World War. Miller served as radioman and tail gunner in the US Army Air Corps in 53 bombing missions over Italy, including the destruction of the Benedictine Abbey at Monte Cassino. The trauma changed his life. Though Miller stayed a Catholic, he was constantly in conflict with the Church. After writing *A Canticle for Leibowitz*, he became a

Figure 4.4 The Catholic Order of Leibowitz, named after a twentieth-century physicist who bestows upon humanity the task of preserving pre-holocaust knowledge. Original artwork by Peter Thorpe.

recluse. He shot himself in 1996 after writing the greater part of a sequel to his one and only classic.

Flame Deluge

Canticle is set in an abbey in post-holocaust America. Originally published in three parts, the plot unravels during three ages (Fiat Homo, Fiat Lux, Fiat Voluntas Tua) roughly six hundred years apart, as civilization re-emerges from an apocalypse. The novel's atmosphere springs from its richly evocative setting in the south-western desert of the USA. Its narrative fuses dark comedy with a sober examination of the questions of faith, science, and power.

After the Flame Deluge destroys civilization in the twentieth century, the Age of Simplification begins. It is the twenty-sixth century. A violent backlash rages against the culture of science that led to nuclear war. Those of learning are slaughtered. Literacy becomes near extinct, and books are destroyed *en masse*. In this way, the first part of the novel ('Fiat Homo: Let There Be Man') paints a new Dark Age.

Echoing the times after the fall of the Roman Empire, scraps of pre-holocaust knowledge, the Memorabilia, are collected and conserved by remote communities of Catholic priests. They understand little of the ancient manuscripts of which they are the keepers. Aptly, the first nuclear war is cast in the form of a biblical narrative, 'God, in order to test mankind... had commanded the wise men of the age... to devise great engines of war... weapons of such might that they contained the very fires of Hell... and... suffered these magi to place the weapons in the hands of princes'[72].

The princes, failing to recognize mutually assured destruction, ignore the advice of the wise men, 'If I but strike quickly enough, and in secret, I shall destroy those others in their sleep, and there will be none to fight back; the Earth shall be mine. Such was the folly of princes, and there followed the Flame Deluge'[73].

The Memorabilia remain a mystery until around 3174, when a new Renaissance begins. Thon Taddeo, a secular scholar, is the intellectual of the age. He is compared to that scarcely recalled sage before the Flame Deluge, Albert Einstein. Through exacting study of the Memorabilia, Thon Taddeo deciphers a new science out of the old. From this revolution, energy is again created ('Fiat Lux: Let There Be Light').

By the last age of *Canticle* ('Fiat Voluntas Tua: Thy Will Be Done'), the year 3781, mankind has starship technology. Such progress has, however, again led to the making of two superpowers, the Asian Coalition and the Atlantic Confederacy. Once more the two sides are engaged in a Cold War, both with hydrogen weapons. Finally, as nuclear weapons are assembled in space, a starship of the church blasts off as a new Flame Deluge begins. The last monk to board, knocking the dirt from his sandals, speaks softly, 'Sic transit mundus' (thus passes the world). A far more damning indictment than the traditional 'sic transit gloria mundi' (thus passes the glory of the world).

Fiat Homo: Fiat Lux: Fiat Voluntas Tua

Most science fiction is sceptical of religion, on the rare occasion when spiritual issues are considered. *Canticle*, however, is notable for the power of its pious reflections, though it often strays from convention. The central paradox to Miller's book is this: the more humans know about science, the less they know about themselves.

At first ('Fiat Homo') little is understood about the scientific secrets of the Memorabilia. During the Age of Simplification, the monks not only preserve the ancient knowledge, they also attempt to control and rewrite history. As in Orwell's *Nineteen Eighty-Four*, he 'who controls the past controls the future: who controls the present controls the past'[74]. Once science reasserts itself ('Fiat Lux'), scholar Thon Taddeo is sceptical of the church's account of history. Man is once more the measure of all things, and a new age of enlightenment begins. Taddeo is aware of the military ambitions of his monarch, and yet he absolves himself of all responsibility:

> He had a choice: to approve of them, to disapprove of them, or to regard them as impersonal phenomena beyond his control like a flood, famine, or whirlwind. Evidently, then, he accepted them as inevitable – to avoid having to make a moral judgement[75].

The pride and arrogance of science, to 'be as gods'[76], will soon lead to the same tragic error of the Bomb culture. To underline the point, Miller later labels the Bomb 'Lucifer' (light-bearer), but he is careful not to set religion against science. Just as the Jesuits of the Renaissance advanced the ideas of Copernicus in China and Japan, *Canticle*'s monks first preserve ancient science as an integral part of the divine.

As the final starship blasts off from Earth as 'the visage of Lucifer mushroomed into the hideousness above the cloudbank'[77], the brothers once again become the custodians of the Memorabilia, 'It was no curse, this knowledge, unless perverted by Man'[78]. Apocalyptic visions flourish in the last section of *Canticle* ('Fiat Voluntas Tua'). Nuclear holocaust augurs the end of time as foretold in the Book of Revelations. Miller utters the novel's overwhelming question:

> Listen, are we helpless?... doomed to do it again and again? Have we no choice but to play phoenix in an unending sequence of rise and fall?... Egypt, Greece... Rome... Britain, America – burned into the oblivion of the centuries.... Are we doomed to do it, Lord, chained to the pendulum of our own mad clockwork, helpless to halt its swing?[79]

The closing nuclear blast would seem to suggest a dark future. The cycle of annihilation seems doomed to recur until human history is eradicated.

In 1954, British philosopher of history Herbert Butterfield had suggested:

> If men put their faith in science, and make it the be-all and end-all of life, as though it were not to be subdued to any higher ethical end, there is something in the very composition of the universe that will make it execute judgement on itself, if only in the shape of the atomic bomb[80].

Miller offered, five years later in *Canticle*, that human understanding, in the form of science without responsibility, remains fragmentary, 'until someday – someday, or some century – an Integrator would come, and things would be fitted together again'[81].

Yet there is promise of resurrection. The monks who leave the dying planet on a starship take with them a few chosen starship voyagers. But they also carry the microfilmed Memorabilia. God essentially liberates this record of the secular and sacred human history, literature and science. The space flight acts as a providential sign that the human race, if not the planet, will continue[82].

Cat's Cradle

A Canticle for Leibowitz was a dire warning. Man will only continue to play a part in history by waging war on his innate evil. Responsible, moral actions of scientists trigger a series of consequences, for good or ill. The Manhattan Project was viewed as the ultimate abdication of such moral responsibility, of which scientists were directly accused. True, a minority of scientists were strongly motivated to communicate the dangers of nuclear war. Some, like Leo Szilárd, even dedicated the rest of their lives to this effort.

So too did Kurt Vonnegut. Like Walter M. Miller Jr, the great American novelist was emotionally traumatized by the atrocities of war. Vonnegut was left physically unscathed after the wanton British firebombing of the city of Dresden in 1945. Witness to the senseless slaughter of 135,000 people, he spent a quarter of a century coming to terms with such outrageous carnage. It was the largest massacre in European history[83]. Vonnegut finally faced his Dresden demons in his late sixties novel, Slaughterhouse Five, named after the underground meatpacking cellar that saved his skin.

To realize the full horror of his experience, Vonnegut had turned to science fiction. He was on the verge of abandoning writing altogether when one of his early novels became a bestseller. The book was Vonnegut's Bomb classic, Cat's Cradle (1963). The novel confronts the increasing possibility of our wiping out the world through human folly. It critically questions the deification of science and technology[84]. Together with works such as Slaughterhouse Five, Vonnegut's fiction is widely regarded as one of the most significant contributions to American literature in the twentieth century.

Cat's Cradle is a mock-apocalyptic novel. It satirizes doomsday books such as *On The Beach* and *A Canticle for Leibowitz*. It also confronts the biggest topic possible: man's potential self-destruction through physics. The story's narrator is busy gathering information for a book to be called *The Day the World Ended*. This book within a book is to be an account of what notable Americans had done on 6 August, 1945, 'the day when the first atomic bomb was dropped on Hiroshima, Japan'[85]. Whereas many critics had sought to blame the contemporary nuclear dilemma on unrestrained technology, Vonnegut begged to differ. It is the failure to be fully human that is especially dangerous.

The main butt of Vonnegut's satire is science. Felix Hoenikker, although dead, is the central character of the story. Famously branded 'the father of the atom bomb', the fictional Hoenikker is a quirky fusion of real-life H-bomb pioneers Edward Teller and Robert Oppenheimer. The narrator focuses on the human not the technical side of the Bomb. He finds in Hoenikker a social imbecile, a solipsistic 'genius' who cares nothing for the applications of his research. Indeed, Hoenikker ends up destroying life on Earth through his creation of *ice-nine*, an alternative structure of water he originally discovers as a mental puzzle, after a casual chat with a Pentagon general.

On his last visit to Los Alamos, Austrian journalist, Robert Jungk, had told of his encounter with a mathematician:

His face was wreathed in a smile of almost angelic beauty. He looked as if his gaze was fixed upon the world of harmonies. But in fact he told me later that he was thinking about a mathematical problem whose solution was essential to the construction of a new type of H-Bomb[86].

For this true-life scientist, who never observed a single explosion of the bombs he helped detonate, 'research for nuclear weapons was just pure mathematics, untrammelled by blood, poison or destruction'[87]. As Italian physicist Enrico Fermi famously said,

'Don't bother me about your scruples. After all, the thing is beautiful physics'[88].

Similarly, for Felix Hoenikker, research into weapons of mass destruction is pure play. It is this distorted innocence that hastens his discovery of *ice-nine*, a substance that proves to be even more apocalyptic than the Bomb itself. In his darkly funny way, Vonnegut takes this innocence one step further in his portrayal of Hoenikker family life. Angela Hoenikker, Felix's daughter, felt she had three children:

> me [Newt], Frank and Father. She wasn't exaggerating, either. I can remember cold mornings when Frank, Father and I would be all in a line in the front hall, and Angela would be bundling us up, treating us exactly the same. Only I was going to kindergarten; Frank was going to junior high; and Father was going to work on the atomic bomb[89].

Later, Felix Hoenikker becomes so absorbed with turtles that he stops working on the Bomb. Anxious Manhattan Project officials pay a visit to Angela, desperate for advice. Take away the turtles, is her simple solution. Indeed, the mind of this eminent research physicist is so puerile that he 'came to work the next day and looked for things to play with and think about, and everything there was to play with and think about had something to do with the atom bomb'[90].

The glitter of nuclear weapons

Vonnegut is no neo-Luddite. Rather, *Cat's Cradle* is a comic portrayal of a scientist stripped of *all* moral responsibility. It is a spiky caveat of the consequences of the veneration of science and technology. Vonnegut marks the limits of the dream of scientific progress. *Cat's Cradle* does not suggest we abandon science. But it rails against the nuclear state. For Vonnegut the antinuclear struggle is a fight for human power, a war waged against technocracy.

English physicist Freeman Dyson had suggested that 'scientists rather than generals took the initiative in getting nuclear weapons programs started'[91], and that they were 'motivated to build weapons by feelings of professional pride as well as of patriotic duty'[92] rather than strategic needs.

Vonnegut's novel presents the contrasting options that faced such scientists. After a test detonation of the Bomb, a physicist turns to Felix Hoenikker and says 'science has now known sin', a clear reference to the celebrated Oppenheimer quote. Hoenikker's ingenuous response was to ask, 'What is sin?'[93]. In great contrast, once the Bomb is dropped on Hiroshima, another scientist announces he is quitting because 'anything a scientist worked on was sure to wind up as a weapon, one way or another'[94], and that he 'didn't want to help politicians with their fugging wars anymore'[95].

Felix Hoenikker is a cultural Neanderthal. Sealed off in his research lab, and never having read 'a novel or a short story in his life'[96], he simply cannot imagine the brunt of his 'technically sweet' dabbling on flesh and blood humans. In contrast, Robert Oppenheimer was famed for his erudition and his intimate knowledge of literature and scripture. His collusion in the Bomb culture is even more disturbing. 'If scientists as sensitive as Oppenheimer can indeed wall off their moral sensibilities so completely and successfully', wrote Philip M. Stern, 'then technology is an even more fearsome monster than most of us realise'[97].

Indeed, in Oppenheimer's own words, 'when you see something that is technically sweet you go ahead and do it... that is the way it was with the atomic bomb. I do not think anybody opposed making it; there were some debates about what to do with it after it was made'[98]. Polish-British physicist and nuclear sceptic Joseph Rotblat agreed: 'scientists felt that only after the test... should they enter into the debate about the use of the Bomb'[99]. Freeman Dyson acknowledged the draw: 'I have felt it myself, the glitter of nuclear weapons. It is irresistible if you come to them as a scientist. To feel its there in your hands – to release this energy that fuels the stars – to lift a million tons of rock into the sky', he felt was, 'partly responsible for all our troubles'[100].

Cat's Cradle drops a fictional H-bomb on technocracy. The novel also blasts the tendency to make mindless Felix Hoenikkers expedient scapegoats[101]. In condemning the complicity of physicists and politicians, readers may fail to recognize their own responsibility. Vonnegut understands there are no experts on nuclear war. No real experts on strategy, tactics or deterrence. And no experts on winning nuclear wars. Only by becoming engaged in the debate can we act as responsible humans in the nuclear age.

Killing joke

After Sputnik, Cold War tensions hit their nuclear peak. Amidst a heated debate on the allegedly widening 'missile gap' between the US and the Soviet Union, came the Cuban Missile Crisis. For a couple of weeks in October 1962, the two superpowers stood on the brink of war. The rest of the world looked on, horrified and helpless. It was the most dangerous fortnight in human history. With both sides stunned in the aftermath of the crisis, Kennedy initiated the hotline communication that still sits between Washington and Moscow. Mutually assured destruction was a wake-up call for the world. Not for everyone, it seems.

A renegade band of physicists went looking for a quick nuclear fix. The consensus that a nuclear war was inherently unwinnable was illogical to these scientists-turned-strategists. As with the survivalist narratives and chauvinistic fiction of Pat Frank and Robert Heinlein, the physicists estimated how many lives the US could lose, yet still rebuild.

RAND Corporation theorist Herman Kahn was one of those who argued 'the unthinkable' – a pre-emptive nuclear strike against the Soviet Union. Like Felix Hoenikker, to Kahn it was all play. The strategic method for his madness was game theory. Kahn's plan rested upon two highly contentious ideas. First, that nuclear war was likely. Second, that like any other war, it was winnable.

Figure 4.5 Peter Sellers as Dr Strangelove, the sinister German title-character – an amalgamation of RAND Corporation strategist Herman Kahn, Nazi SS officer-turned-NASA rocket scientist Wernher von Braun, and 'father of the hydrogen bomb' Edward Teller.

The fragile nature of the Cold War conflict and the doctrine of mutual assured destruction were ripe for satire. In 1964, the subtle wit of *A Canticle for Leibowitz* and the sharpened black humour of *Cat's Cradle* were raised to an art form by Kubrick's dark comic masterpiece, *Dr Strangelove* (subtitled *or How I learned to stop worrying and love the Bomb*).

The world premiere was originally scheduled for late 1963, but was delayed until January 1964 in light of the assassination of John F. Kennedy. The Cuban Missile Crisis was still fresh in the public psyche. It was the movie's undoubted brilliance that led to its huge impact, however.

Dr Strangelove earned four Academy Award nominations, including best picture, director, and screenplay for Kubrick, and best actor for Peter Sellers. Famous film critic Roger Ebert suggests that *Dr Strangelove* arguably remains the best political satire of the century. The film was listed at number 26 on the American

Film Institute's *100 Years, 100 Movies* and at number 3 on the AFI's *100 Years, 100 Laughs*. After all, what could be more absurd than the very idea of two mega-powers willing to wipe out all human life on the planet.

Kubrick had waged cinematic war once before. His 1957 humanitarian classic, *Paths of Glory*, was an anti-war movie. Staged on the Great War battlefields of France, it featured clashes between rulers and the ruled. Kubrick was now rapt by the notion of nuclear blunder. He felt personally vulnerable[102].

His motivation was the 1958 Cold War thriller *Red Alert* by Welsh author Peter George. The book was a grim warning. Its caveat was the absurd ease with which nuclear apocalypse might be accidentally triggered. *Red Alert* was serious melodrama, and George a rather sombre ex-RAF navigator who had recently joined the Campaign for Nuclear Disarmament. Kubrick viewed George's idea of chance catastrophe as too farcical for drama. It was a killing joke. In Kubrick's words, 'How the hell could the President ever tell the Russian Premier to shoot down American planes? Good Lord, it sounds ridiculous'[103].

So *Red Alert* repeated itself, first as tragedy, second as farce. *Dr Strangelove* was born. The story became a high-powered political lampoon. The movie has three basic backdrops – an airbase, the cockpit of a B-52 and a War Room beneath the Pentagon. The grotesque cartoon characters that inhabit these settings become more frenzied as the crisis spirals. *Dr Strangelove* is a dark satire of the military-industrial power first anticipated by Wells in *The World Set Free*, and set in a war room first prototyped in the science fiction of E. E. 'Doc' Smith.

Jack D. Ripper is the rabidly paranoid and maverick Strategic Air Command General at the airbase who believes the communists have fluoridized the US water supply to make him impotent. He orders an attack on the Soviets that cannot be recalled. Air Force Major 'King' Kong is the cowboy captain of the rogue B-52 bomber, the lone plane that gets through Soviet defences. General Buck Turgidson, a strategic bombing enthusiast, and the bald

President Merkin Muffley are among the grotesques who desperately try to salvage peace in the claustrophobic War Room. Then, of course, there is Strangelove himself.

A strange love revealed

Strangelove is a black comic amalgam of sinister science. He is one part Herman Kahn, one part Wernher von Braun, and two parts Edward Teller. Kubrick had met Kahn, and became engrossed in Kahn's controversial book *On Thermonuclear War*. Kahn had proposed the 'logical' notion of winnable nuclear warfare. Kahn gave Kubrick the idea for a fictional Doomsday Machine, capable of wiping out the entire planet in the event of a nuclear attack. Wernher von Braun gave Strangelove that Teutonic touch. Peter George's novel of the screenplay refers to Strangelove's black-gloved hand, which was a memento of his time 'working on the German V-2 rocket'[104].

Edward Teller was a true-life Strangelove. Father of the superbomb, and the classic scientist turned strategist, Teller was the apocalyptic nuclear 'sage' of the Cold War. He was infamously obsessed with security. He was the only member of the scientific community to smear Oppenheimer as a security risk during Oppenheimer's trial. Teller's hare-brained solutions invariably used hydrogen fusion weapons, and he held extreme opinions on the Red Menace of the Soviets. Teller even had a disability. His lower leg was severed in a Munich tram accident in his youth, requiring him to wear a prosthetic foot, leaving him with a life-long limp.

Dr Strangelove is one of the most influential archetypes of the scientist in cinema[105]. He is mad, in the custom of Victor Frankenstein; he is prosthetic, implying machine-like inhumanity; and he is corporate, detached from any personal responsibility through the collective cover of the 'Bland Corporation', an obvious reference to Kahn's position at the RAND Corporation. His black-gloved arm forever threatens to expose Strangelove's mania for destruction, jerking into a 'Sieg Heil' at the President with erratic and embarrassing zeal. As Peter Sellers put it, 'the arm hated the rest of the body for having made a compromise – that arm was a Nazi'[106].

In the movie Strangelove first enters the fray from the shadows. Vonnegut had focused on the puerile complicity of nuclear physicists. Kubrick's parody is more damning still. His target is the scheming malevolence of Teller, and the unrelenting rationality behind Kahn's military strategy. Cold, calculated reason ultimately ends in holocaust.

As the President utters the illustrious line to the bickering General Turgidson and the Russian Ambassador, 'Gentlemen, you can't fight in here. This is the War Room', Strangelove prepares to deliver his master plan. With unerring logic he talks of preserving a 'nucleus of carefully selected specimens' of the human race at the bottom of America's deeper mineshafts for a hundred years, 'After all, the conditions would be far superior to those of the *so-called* concentration camps, where there is ample evidence most of the wretched creatures clung desperately to life'.

When a disbelieving President, the only wise man in the Pentagon, questions the immensity of the task, Strangelove's plan is unveiled:

Strangelove: It would not be difficult, Mein Führer. Nuclear reactors could – I'm sorry, Mr President – nuclear reactors could provide power almost indefinitely; greenhouses could maintain plant life; animals could be bred and slaughtered...

President: I would hate to have to decide who stays up and who goes down.

Strangelove: That would not be necessary, Mr President; it could easily be accomplished with a computer; and the computer could be set and programmed to accept factors from youth, health, sexual fertility, intelligence and a cross-section of necessary skills. Of course it would be absolutely vital that our top government and military men be included, to foster and impart the required principles of leadership and tradition...

And on this technocratic thought, Strangelove's Nazi arm thrusts into another 'Sieg Heil'.

The master plan of Strangelove is that of scientist as god, re-colonizing the Earth in his own image[107]. Wells' World State is taken to its farcical and fascist conclusion. Strangelove draws prodigious strength from the holocaust raging outside the War Room. As Major 'King' Kong rides his nuclear charge like a phallic rodeo mount, and the mushrooming orgasm of global apocalypse begins, Strangelove rises out of his wheelchair and yells, 'Mein Führer – I can walk!'. This outrageous parody of an erection reveals the motive for the movie's title – the scientist's strange erotic love of mass destruction.

Surfeit of monsters

Stanley Kubrick was praised by American historian of science, Lewis Mumford, for placing Strangelove right at the centre of his nightmare. It is, after all, where he belongs. So potent was the film's imagery that President Ronald Regan, on first entering office, is said to have asked for the way to the Pentagon War Room. As internationally renowned social critic Theodore Roszak suggests:

Modern science provides us with a surfeit of monsters, does it not? I realise there are many scientists – perhaps the majority of them – who believe that these and a thousand other perversions of their genius had been laid unjustly at their doorstep. These monsters, they would insist, are the bastards of technology: sins of applied, not pure science'[108].

Roszak is unconvinced by attempts to blame everything on technology and politicians, 'Perhaps it comforts their conscience somewhat to invoke this much-muddled division of labour...

Dr Faustus, Dr Frankenstein, Dr Moreau, Dr Jekyll... Dr Strangelove. The scientist who does not face up to the warning in this persistent folklore of mad doctors is himself the worst enemy of science'[109].

It is these images from science fiction that reflect a deep-seated public fear of science, a fear of the arrogance of science and scientists. There was arrogance about the reduced and inhuman notion of knowledge; arrogance about the advising of politicians; and arrogance about turning atrocities into statistics. The Cold War cynicism over the politics of government and the fallout of the culture of science was no longer enough. It was time for change.

References

1. Democritus quoted in White, M. (2001) *Rivals*. London.
2. Chamberlin, T. C. (1899) *Science*, **9**, 12.
3. Walters, P. and Hey, T. (1997) *Einstein's Mirror*. Cambridge University Press, Cambridge, p. 135.
4. *Ibid*., p. 135.
5. Wells, H. G. (1914) *The World Set Free: A Story of Mankind*. Macmillan, New York.
6. *Ibid*., p. 44.
7. *Ibid*., p. 286.
8. *Ibid*., p. 178.
9. *Ibid*., p. 117.
10. Rhodes, R. (1986) *The Making of the Atomic Bomb*. Simon & Schuster, New York, p. 26.
11. *Ibid*., p. 27.
12. Burden, C. N. (2003) *Nuclear Paranoia*. Pocket Books, London, p. 19.
13. De Ropp, R. S. (1972) *The New Prometheans*. London.
14. Dowling, D. H. (1986) The atomic scientist: machine or moralist? *Science Fiction Studies*, #39, **13**(2), p. 139.
15. *Ibid*., p. 140.
16. *Ibid*., p. 140.
17. Lucie Donahue: Blast script on Eutopias.
18. Orwell, G. (1946) Letter to the Reverend Herbert Rogers. In: *Collected Essays, Journalism and Letters Volume 4*. Penguin, Harmondsworth, p. 129.

19. Orwell, G. (1985) *The War Broadcasts* (ed. W. J. West). Marlboro, London, p. 51.
20. Orwell, G. (1937) *The Road to Wigan Pier*. Penguin, London, p. 226.
21. Orwell, G. (1941) Wells, Hitler and the World State. In *Collected Essays, Journalism and Letters*, Vol. 2. Penguin, Harmondsworth, p. 171.
22. Orwell, G. (1941) *Unwelcome Guerilla: George Orwell and the New Statesman – An Anthology*. London, p. 42.
23. Lucie Donahue: Blast script on Eutopias.
24. Orwell, G. (1942) The rediscovery of Europe. In: *Collected Essays, Journalism and Letters*, Vol. 2. Penguin, Harmondsworth, p. 234.
25. Orwell, G. (1940) Charles Dickens. In: *Collected Essays, Journalism and Letters*, Vol. 1. Penguin, Harmondsworth, p. 488.
26. Orwell, G. (1941) Wells, Hitler and the World State. In: *Collected Essays, Journalism and Letters*, Vol. 2. Penguin, Harmondsworth, p. 170.
27. *Ibid.*, p. 167.
28. Orwell, G. (1942) Wartime diary 27 March 1942. In: *Collected Essays, Journalism and Letters*, Vol. 2. Penguin, Harmondsworth, p. 469.
29. Williams, R. (ed.) (1974) *George Orwell: A Collection of Critical Essays*. Prentice Hall, New Jersey.
30. Orwell, G. (1949) *Nineteen Eighty-Four*. Secker & Warburg, London.
31. Orwell, G. (1940) *Collected Essays, Journalism and Letters*, Vol. 1. Penguin, Harmondsworth, p. 331.
32. Orwell, G. (1949) *Nineteen Eighty-Four*. Secker & Warburg, London.
33. Orwell, G. (1946) *Tribune*, 4 January 1946. In: *Collected Essays, Journalism and Letters*, Vol. 4. Penguin, Harmondsworth, pp. 95–9.
34. Orwell, G. (1937) *The Road to Wigan Pier*. Penguin, London, p. 225.
35. Orwell, G. (1942) *Collected Essays, Journalism and Letters*, Vol. 2. Penguin, Harmondsworth, pp. 32–3.
36. Orwell, G. (1945) *Observer*, 14 January 1945. In: *Collected Essays, Journalism and Letters*, Vol. 3. Penguin, Harmondsworth, pp. 364–5.
37. Orwell, G. (1944) *Tribune*, 4 February 1944. In: *Collected Essays, Journalism and Letters*, Vol. 3. Penguin, Harmondsworth, p. 110.
38. Seed, D. (1999) *American Science Fiction and the Cold War*. Edinburgh University Press, Edinburgh, p. 68.

39. *Ibid.*, p. 69.
40. Riesman, D. (1964) *Abundance for What? And Other Essays*. Chatto & Windus, London.
41. Orwell, G. (1942) *Collected Essays, Journalism and Letters*, Vol. 2. Penguin, Harmondsworth, pp. 32–3.
42. Deutscher, I. (1954) 1984 – The mysticism of cruelty. In: *Heretics and Renegades*. Hamish Hamilton, London.
43. *Ibid.*
44. Orwell, G. (1949) *Collected Essays, Journalism and Letters*, Vol. 4. Penguin, Harmondsworth, p. 564.
45. *Ibid.*
46. Aldiss, B. (1986) *Trillion Year Spree*. Victor Gollancz, London.
47. Shelley, M. (1818) *Frankenstein: or, The Modern Prometheus*, London.
48. Franklin, H. B. (1986) Strange scenarios: science fiction, the theory of alienation, and the nuclear gods. *Science Fiction Studies*, #39, 13(2), p. 117.
49. Stewart, G. R. (1949) *Earth Abides*. Greenwich, p. 18.
50. *Ibid.*, p. 12.
51. Walters, P. and Hey, T. (1997) *Einstein's Mirror*. Cambridge University Press, Cambridge, p. 134.
52. Burden, C. N. (2003) *Nuclear Paranoia*. Pocket Books, London, p. 20.
53. Rhodes, R. (1986) *The Making of the Atomic Bomb*. Simon & Schuster, New York, p. 749.
54. *Ibid.*, p. 751.
55. Burden, C. N. (2003) *Nuclear Paranoia*. Pocket Books, London, p. 44.
56. Clute, J. and Nicholls, P. (1999) *The Encyclopedia of Science Fiction*. Orbit, London, p. 890.
57. Newman, K. (1999) *Millennium Movies*. Titan, London.
58. Shaheen, J. G. (1978) *Nuclear War Films*. Southern Illinois University Press, Carbondale, IL.
59. Burden, C. N. (2003) *Nuclear Paranoia*. Pocket Books, London, p. 44.
60. Seed, D. (1999) *American Science Fiction and the Cold War*, Edinburgh, p. 169.
61. Teller, E. (1962) *The Legacy of Hiroshima*, MacMillan, p. 241.
62. Miller, R. (2006) *Science Communication in a Time of War(s) – World and Cold*, Glamorgan.
63. Seed, D. (1999) *American Science Fiction and the Cold War*. Edinburgh University Press, Edinburgh, p. 169.

64. Weart, S. R. (1988) *Nuclear Fear: A History of Images*. Harvard University Press, Cambridge, MA, p. 219.
65. Kearns, D. (1976) *Lyndon Johnson and the American Dream*. New American Library, New York, p. 145.
66. Editorial, *Manchester Guardian*, 7 October 1957.
67. Seed, D. (1999) *American Science Fiction and the Cold War*. Edinburgh University Press, Edinburgh, p. 63.
68. Clarkson, H. (1959) *The Last Day*. New York, p. 123.
69. *Ibid.*, p. 44.
70. *Ibid.*, p. 37.
71. Seed, D. (1999) *American Science Fiction and the Cold War*. Edinburgh University Press, Edinburgh, p. 63.
72. Miller, W. M. (1959) *A Canticle for Leibowitz*. Lippincott, New York, p. 57.
73. *Ibid.*, p. 57.
74. Orwell, G. (1949) *Nineteen Eighty-Four*. Secker & Warburg, London, p. 214.
75. Miller, W. M. (1959) *A Canticle for Leibowitz*. Lippincott, New York, p. 198.
76. *Ibid.*, p. 220.
77. *Ibid.*, p. 312.
78. *Ibid.*, p. 265.
79. *Ibid.*, p. 245.
80. Butterfield, H. (1954) *Christianity and History*. Bell, London, p. 60.
81. *Ibid.*, p. 62.
82. Manganiello, D. (1986) History as judgement and promise in *A Canticle for Leibowitz*. *Science Fiction Studies*, #39, **13**, p. 159.
83. Vonnegut, K. (2006) *A Man Without a Country*. Bloomsbury, London, p. 17.
84. Zins, D. (1986) Rescuing Science from Technocracy: *Cat's Cradle* and the play of apocalypse. *Science Fiction Studies*, **13**, 170.
85. Vonnegut, K. (1963) *Cat's Cradle*. Penguin, London, p. 7.
86. Jungk, R. (1958) *Brighter Than a Thousand Suns: A Personal History of the Atomic Scientists*. Harcourt, New York, p. 208.
87. *Ibid.*, p. 209.
88. Quoted in Zins, D. (1986) Rescuing Science from Technocracy: *Cat's Cradle* and the play of apocalypse. *Science Fiction Studies*, **13**, 170.
89. Vonnegut, K. (1963) *Cat's Cradle*. Penguin, London, p. 19.
90. *Ibid.*, p. 20.

91. Dyson, F. (1984) Weapons and hope, Part II. *The New Yorker*, 13 February, pp. 67–117.
92. *Ibid.*
93. Vonnegut, K. (1963) *Cat's Cradle*. Penguin, London, p. 21.
94. *Ibid.*, p. 27.
95. *Ibid.*, p. 27.
96. *Ibid.*, p. 16.
97. United States Atomic Energy Commission (USAEC) (1971) *In the Matter of J Robert Oppenheimer: Transcript of Hearing Before Personal Security Board and Text of Principal Documents and Letters*. Cambridge.
98. Erikson, K. (1985) Reflections on the Bomb: of accidental judgements and casual slaughters. *The Nation*, 3–10 August, p. 60.
99. Rotblat, J. (1985) Leaving the bomb project. *Bulletin of the Atomic Scientists*, **41**, 18.
100. Weart, S. R. (1982) The day after Trinity. *Bulletin of the Atomic Scientists*, **38**, 42.
101. Zins, D. (1986) Rescuing Science from Technocracy: *Cat's Cradle* and the play of apocalypse. *Science Fiction Studies*, **13**, p. 175.
102. Frayling, C. (2005) *Mad, Bad and Dangerous?* Reaktion Books, London, p. 102.
103. *Ibid.*, p. 103.
104. George, P. (1963) *Dr Strangelove*. Transworld, London, p. 34.
105. Frayling, C. (2005) *Mad, Bad and Dangerous?* Reaktion Books, London, p. 105.
106. *Ibid.*, p. 103.
107. *Ibid.*, p. 107.
108. Roszak, T. (1974) The monster and the Titan: science, knowledge and gnosis. *Daedalus*, **103**(3), 17–32.
109. *Ibid.*, p. 17–32.

Chapter 5

STRANGER IN A STRANGE LAND: THE NEW AGE

On 20 August 1977, mankind launched its most expensive message in a bottle. The $250 million, 815 kilogram unmanned Voyager 2 probe blasted off atop a Titan III-E Centaur rocket bound for the outer solar system. Voyager sits as the iconic representation of the New Age's desire to both explore and unite, to experiment and to limit. The most prolific of our deep space objects, it was part of the New Age's attempts to explore outer space. Symbolically it held within its metal body an exploration of innerspace – an attempt to chart the human condition rendered permanent on two golden discs[1]. These were an extension of the interstellar communication first created for the plaques of Pioneer 10 and 11[2]. The golden discs constrained the entirety of our existence into a small sample of sounds, music, and images. These were chosen by a team led by the American astronomer, astrobiologist, influential science popularizer and science fiction author Carl Sagan. An archive of 115 pictures, spoken greetings in 55 languages, a selection of natural and man-made sounds, and a 90 minute collection of music from around the world. They became an interstellar Rosetta stone. Their true purpose was not to spend their life wandering their errant way across the galaxy to be discovered by some distant intelligence. Their true message was not aimed at the stars, but back towards the Earth. In a single artefact, the Voyager 2 mission tried to embrace the whole globe in one endeavour. It tried to reunite the fractured society it left behind. A world divided in politics, science and culture.

Moon shot

For the first time we start to see not just the intersection of individual cultural, scientific and social perspectives but their fusion to create something new and challenging. The Voyager probes were the logical realization of the foresight of the Astounding Age and of their incarnation in the 'space race' of the 1960s – science fictional prophecies which said that mankind would reach out to the stars aboard rocket ships, and that we would walk upon new worlds and gain new perspectives on what it meant to be human. That we would boldly go where no man had gone before[3]. These insights were incarnated in the 'space race', the military-industrial struggle between the United States and the Soviet Union, that began after the Soviet launch of Sputnik 1 on 4 October 1957 and involved US and Russian efforts to explore outer space with artificial satellites, to send humans into space, and to land people on the Moon.

Figure 5.1 Man on the Moon (© NASA).

The drive from science fiction to place mankind into space involved the synthesis of many different aspects of science. Never before had such disparate scientific elements been called to work together in a united effort. The Apollo Moon missions erased the divisions between the sciences and fused them into one endeavour. Mankind's efforts to divide its knowledge into component parts had to be re-examined in the light of the interrelations which the space race provided. A nexial[4] approach which emphasized the cross-disciplinary nature of the acquisition of knowledge moved to the forefront in the New Age.

Heat death – Pamela Zoline

A locus for this movement can be found in one of the most overlooked works of science fiction to emerge within the New Age, Pamela Zoline's 1967 Heat Death of the Universe, a story which consists of 54 numbered paragraphs describing the domestic drudgery of Sarah Boyle, a Californian 'mom', as she prepares for her child's birthday. Interjected into this familial account are reflections on entropy, chaos and the heat death of the Universe. Written by the polymath Zoline for Michael Moorcock's New Worlds magazine, this short story was not only the first she had written since leaving school, but also appeared in the same month that Zoline had one of her paintings exhibited in the Tate Gallery[5].

It may not seem that Heat Death is a science fiction story. Its incongruity, experimentalism and political overtones mark it out as a classic example of the new wave of science fiction, a wave that was breaking over the shores of what was, up until this point, a fairly conservative genre. Moorcock, the editor of New Worlds, comments that in her short story Zoline connects 'the modern myths of science (entropy, etc.) as they are understood by the layman with that great myth figure of modern fiction, the Victimized Domestic Woman'[6]. Heat Death subverts the domestic ideal. It juxtaposes the gradual cosmic exhaustion of the entire Universe with the mundane. Zoline's story, like much of the New Age, embraces new patterns of thinking, feeling and behaving.

Some have attempted to place Zoline's work within a traditional science fiction mode. It could be an alien invasion text, with entropy as an outside force invading the like of the housewife Boyle[7]. Or an apocalyptic tale[8], a hangover from the motifs of the Atomic Age. Instead Zoline's story refuses to be confined by previous definitions of science fiction[9]. It allows the science of physical cosmology to find abstract expression in the medium of fiction. Its inability to be clearly demarcated is an indication of the innovative and iconoclastic approach to science fiction that was to develop during the course of the New Age.

Turn on, tune in, drop out – Timothy Leary

New patterns of thinking, feeling and behaving fused together in a New Age counter-culture. Defined by what it was not, this counterculture was conspicuously multi-faceted. It identified itself clearly as having strong bonds with alternative movements in history. It also associated itself with the non-Western cultures which were starting to emigrate around the world. Throughout the New Age the counterculture conceived itself, and was conceived by society, as including many sub-cultures, tendencies and themes. Amongst the most famous of these was the 'Beat' movement made synonymous with the writer William S. Burroughs.

The New Age's counterculture began in the USA as a reaction against the conservative societal structures endemic in the 1950s. Both the political conservatism and social repression of the Cold War period stultified and constricted an entire generation of young men and women. All over America, and then throughout the Western world, students in campuses and young people in communities started to break free from traditional modal thinking. They embraced a new orthodoxy which proclaimed 'Turn on, tune in, drop out'.

This iconic counterculture phrase was coined by the psychologist and LSD guru Timothy Leary in a press conference in New York in September 1966. In this he remarked 'Like every great religion of the past we seek to find the divinity within and to express this revelation in a life of glorification and the worship of God. These ancient goals we define in the metaphor of the present – turn on, tune in, drop out'[10]. Leary later clarified his clarion call by explaining that *turn on* meant stimulating your mind and body. *Tune in* meant living amicably with the world around you. *Drop out* meant renouncing the societal compulsions of school, the military and employment. In Timothy Leary we find the scientific personification of the New Age. As a psychologist, Leary was one of the high priests of this new faith. Studied in the arts of innerspace, it was Leary, his professional colleagues and their disciples who could offer a practical way to achieve nirvana. For them it was the application of science in the forms of pharmacology and psychology which held the answer to the human condition.

The mantra of 'medication, medication, medication' which was adopted at this time came to be associated with one particular drug, LSD. Originally this was a 1938 by-product of research into migraine relief. By the New Age, its adoption by mental health professionals, most notably Harvard psychology professors Drs Timothy Leary and Richard Alper, initially legitimated its use. Both Leary and Alper became convinced of LSD's possible use as a tool for spiritual growth. Their esoteric and contentious investigations asserted a relationship between the LSD experience and the states of enlightenment experienced by many religious practitioners. Ejected by a traditional academic psychology community, Leary and Alper evolved into countercultural spiritual gurus. Their experimentation with psychedelic drugs became a major component of 1960s counterculture, influencing philosophy, art, music, literature and styles of dress.

This type of experimentation was not new. Aldous Huxley, author of the Astounding Age dystopia *Brave New World*, had chronicled his experiences with the mood-altering substance Mescaline in a famous treatise of pharmacology and philosophy, *The Doors Of Perception*. Such was the influence of this pharmacology that in 1962 Huxley, until then most famous for his treatments of the dehumanizing aspects of scientific progress, embraced aspects of utopian thought in his novel *Island*. He commented that in such a utopia 'Science and technology would be used as though, like the Sabbath, they had been made for man, not, (as at present and still more so in the Brave New World) as though man were to be adapted and enslaved to them'[11].

Summer of love – Alfred Kinsey

Science itself had become subverted. The chemical bonds which held these pharmacological factors together became dissolved in a sea of alternatives. The 1967 'summer of love', made popular by the Scott McKenzie song *San Francisco*, brought 75,000 young people from all over the world together. The experimentation which pharmacology encouraged was expressed through a more open and relaxed attitude to sex. This attitude was in part due to the widespread use of pharmacology, but more directly attributed to the relaxation of sexual mores – a relaxation which was the consequence of the publication of the bestselling scientific books of all time. Alfred Kinsey's *Sexual Behaviour in the Human Male* (1948) and its sequel *Sexual Behaviour in the Human Female* (1953) generated a different attitude to sexual relations in the New Age. Just as much as the younger generation of the New Age wanted to embrace each other, so they wanted to reject violence and institutional expressions of repression.

The pacifist anti-war movements which scientists like Szilárd and Einstein embraced following the nightmare visions of the Atomic Age came to be associated with an entire generation. For science fiction the summer of love did not have a profound effect; after all, it had in fact been there right from the very beginning. Science fiction played a pivotal role in redefining sexual mores. It introduced concepts of pacifism and limits to the military application of science. It explored the use of pharmacology and the way that all of these factors relate to wider cultural movements. The science of the New Age drove forward a developing society, but it was science fiction that provided the original catalyst. That this is the case can be seen clearly in a key text of the period: Robert A. Heinlein's 1961 *Stranger in a Strange Land*.

Stranger in a Strange Land – Robert Heinlein

This story is justly famous, and not just for inventing the concept of the waterbed[12]. A bestseller, it is the story of Valentine Michael Smith, the human descendant of an expedition to Mars who, having been raised by the Martians, returns to Earth. *Stranger* attracted a far wider audience than just conventional science fiction readers. Its themes of sexual freedom and liberation, of the search for true meaning and the exploration of the relationship between inner- and outerspace, came to typify the counterculture of the New Age. It is considered by some as the 'bible' of counterculture. The original version, dedicated to science fiction luminary Philip José Farmer, who had pioneered treatment of sexual themes in science fiction, ironically had its most explicit sex scene excised along with another 60,000 words to make it less socially dangerous and more saleable.

It begins unremarkably enough as a straightforward adventure story and only moves beyond these constraints in the latter half of the book. By making Smith wholly unlike mankind in function, but not form, Heinlein is able to explore what it means to be human by gazing upon it through inhuman eyes:

> Mike is our Prometheus – but that's all. Mike keeps emphasizing this. Thou art God, I am God, he is God – all that groks. Mike is a man like the rest of us. A superior man admittedly – a lesser man taught the things the Martians know, might have set himself up as a pipsqueak god. Mike is above that temptation. Prometheus... but that is all[13].

It is because the panorama of its visions combined with the ambiguity of its morality that *Stranger in a Strange Land* became so popular, equally at home in the back pockets of the hardcore science fiction fan (it won a Hugo the year it was published) as in the embroidered knapsack of the hippie. Its diversity became its strength. Such was its influence that the account of the messianic Smith who was persecuted by the large governments and industries of Earth gave birth to religious movements inspired by those of the novel. Their adherents saw immediate parallels between this iconic text and the countercultural movements of the time. Heinlein kept such movements at a distance, reminding fans that he was an agnostic and that his book should be considered a satire on religion.

Published at the beginning of the New Age, Heinlein's novel was to pave the way for more experimental forms of science fiction. Although not as outrageous as some of its successors, it still remains an important text. It explores the balance that needs to be struck between the necessity of public institutions, such as governments, for the peaceful co-existence of men, versus their tendency to corrupt those with whom they come in contact. *Stranger* provided a framework for sustained discussion during the New Age, and is a coherent and measured exploration of how an individual functions in society. Timothy Leary's 'drop out' cries of abstention from any institutional associations were stretched by Heinlein's more mature attempt to ascertain a level of equilibrium between the rights of the individual and those of the society they inhabit.

Nouvelle vague – Jean-Luc Godard

Just as the waves of experimentalism in form and content were beating on the shore of science fiction, so the same wave was swelling in the influential field of cinema. It was during the New Age that iconoclastic film-makers throughout the world started to create their own 'new waves' using film as their medium. In Britain this manifested itself in the working-class movement of the 'angry young men'. Authors such as Kingsley Amis and John Osborne found their work translated onto screen. However, it was the new wave movement in France which turned out to have a lasting influence on science fiction and the science which it drove forward.

Coming from a more innovative perspective, an entire generation of young French *auteurs* would have an influence upon the wider world. Directors like François Truffaut and Jean-Luc Godard turned to science for help in creating the new images which were to characterize their work. Taking advantage of the emergent technology of miniaturization of electronic components, they employed lightweight handheld cameras and lighting rigs to enormous effect. These enabled them to work on location rather than be confined to studio sets. The speed and flexibility which this new technology afforded encouraged their experimentation and improvisation in all aspects – not least in the content.

It was this new style of film, featuring existential themes, which influenced and had so much in common with the science fiction of the time. Up until this point, cinema, like most written work, was trapped by the convention of narrative. New wave *auteurs* broke through the constraints of film-making and started to experiment in form and content. This experimentation, driven by new technological and scientific advances, was to find a similar outlet in the new wave of science fiction writing.

There is a direct link between new wave cinema and new wave science fiction. It was natural that those involved in the new wave of cinema in Europe should look to science fiction as an outlet for their genre-breaking visions. When seeking sources which critiqued contemporary thought, science fiction was a rich seam to be mined. Examples include Truffaut's 1966 film *Fahrenheit 451*, based on the dystopian novel by Ray Bradbury, set in a world where books are forbidden and critical thought is proscribed. Here the central character, Guy Montag, is employed as a 'fireman' whose duty it is to incinerate contravening material at 451 degrees 'the temperature at which book-paper catches fire, and burns...'[14]. In Jean-Luc Godard's 1965 film *Alphaville, une étrange aventure de Lemmy Caution* (Alphaville, a strange adventure of Lemmy Caution) the trenchcoat-adorned hero Lemmy Caution searches the futuristic city of Alphaville for a missing colleague. The metropolis is controlled by Professor von Braun and administered by the Alpha 60 computer system, which executes any who display love, emotion or other non-logical emotions, creating an inhuman and alienating society. Both films demonstrate the power that technology had to control the lives of the individual. In *Alphaville* in particular Godard makes the point that the application of scientific logic, divorced from emotion, can only lead to pain and repression. Science fiction gave these French *auteurs* the arena to explore the limits of scientific, technological and societal progress.

Cut up – Michael Moorcock and Harlan Ellison

The films of the *nouvelle vague* were marked by jump cuts leading the viewer abruptly from one scene to another. So too the new wave of science fiction marked itself out with a set of similar stylistic conventions. J. G. Ballard published an essay in *New Worlds* magazine, one of the pioneering venues for the new wave, in appreciation of beat writer William S. Burroughs. In it he praised his use of the cut-up technique, in which a text is cut up at random and rearranged, in a written parallel to the jump cuts of Godard and Truffaut.

Ballard goes on to utilize this technique in his 1970 text *The Atrocity Exhibition*, a book which features short chapters, each encompassing an entire story, which Brooks Landon comments are condensed novels 'only loosely unified by any recognisable plot progression but... compellingly structured by its obsessive themes and images of a technologically broken world'[15].

In 1964 the newly installed editor of *New Worlds* magazine, Michael Moorcock, set about proclaiming Ballard and Burroughs as the future of science fiction. They were to provide the 'passion, subtlety, irony, original characterisation, original and good style, a sense of involvement in human affairs, colour, density depth, and, on the whole, real feeling from the writer'[16], All of which Moorcock claimed was missing in the conventional writing of the time. Up until Moorcock's tenure *New Worlds* had been, like its transatlantic cousins *Astounding* and *Amazing*, a bastion for *de rigueur* science fiction stories of the like discussed in Chapter 3. Moorcock rang the changes both in format, with a move first to paperback and then to large format magazine, but mostly with content. This content, like the science of its time, discarded the traditional focus of science fiction with its focus on outer space: 'the future envisaged by the science fiction of the 1940s and 1950s is already our past. Its dominant images, not merely of the first Moon flights and interplanetary voyages, but of our changing social and political relationships in a world governed by technology'[17].

At the time, writers like J. G. Ballard, Brian Aldiss and Thomas Disch were seen as part of the general literary mainstream and not necessarily as part of the new wave of science fiction that was breaking down the traditional and conservative pulp science fiction. The pulps' conservatism, of narratives and in the politics of legions of Brylcreemed heroes saving the Universe from destruction – *again*, was directly challenged. Instead, the role of the individual and the nature of human existence, seen so much in the contemporary *nouvelle vague* cinema, is played out on the pages of *New Worlds* magazine. The political commentary associated with Truffaut and Godard can also be seen in the left-wing politics of Moorcock's *New Worlds* and, allied with a reaction against American Conservatism in Harlan Ellison's US anthology *Dangerous Visions*.

Like *New Worlds* in Britain, *Dangerous Visions* championed the short story form of narrative, which was most prominent in the new wave – perhaps because it is difficult to sustain the incisive and audacious style over the longer format of a novel. Whatever the reason, *Dangerous Visions* set out to live up to its title. Hyped by Ellison as a genre-busting revolution, it is an important anthology with plenty of good stories. Some come from old masters, whilst others are from prominent writers of the new wave. Ultimately it fails to live up to its own propaganda. Its importance lies in its attempts to make clear to the wider public that, although their styles and contents may diverge, the confrontational approach of the writers within *Dangerous Visions* remains the same. But just as science fiction was being revolutionized by those short stories, entropic forces were quickly acting upon it, so that it lost its impetus. *Dangerous Visions* was followed by one sequel, but a third has languished unpublished for over thirty years. *New Worlds* saw itself crumble and fall when Norman Spinrad's expletive fuelled *Bug Jack Barron* was serialized over four issues, leading to shops and retailers withdrawing it from the shelves.

Pamela Zoline's *Heat Death of the Universe* was innovative in its use of entropy, the concept that the Universe will irretrievably exhaust itself, as a mirror to examine the same process in human society. Yet that concept of entropy was to apply as much to the new wave itself as to the stories which it created[18]. The experimentation in form and style which marked the beginning of the new wave were to be overtaken by content. Its writers came to use their fictions to pass comment on key socio-cultural issues of the time. The ultimate entropic decay of the new wave was soon at hand.

Heat death and drowned worlds – J. G. Ballard

The concept of entropy was used extensively prior to *New Worlds* and *Dangerous Visions*. In *The Drowned World*, J. G. Ballard's first major novel, published in 1962, Ballard mixes together themes which were to later be explored in the New Age. In a subtle mix of the nature of the individual, the origins of violence and a slow entropic death of civilization we find a new theme – that of ecological disaster.

Ballard's work has a semi-autobiographical note (which was later to become fully realized in his best-known work, *Empire of the Sun*) in that he was born in Shanghai and at the age of seven watched the Japanese invade, remaining with his parents in Shanghai through the Japanese occupation. Witnessing not just the atrocities of daily life with its random violence, Ballard was also later transferred to a prison camp eight miles outside of the city. With its marshy waterways and subtropical inhabitants it acted as a prototype for the oppressive swamps depicted so vividly in *The Drowned World*.

Ecological disaster, fuelled presciently by a runaway greenhouse effect, has resulted in a reversion of landscape and animals to those resembling the Triassic period. Earth's human inhabitants have also retreated into a more primitive form of existence. That the disaster is natural, the consequence of solar storms bombarding the Earth's atmosphere with radiation, is of little comfort to those who remain. Mother Earth has been re-sculpted by immense forces into something both more exotic and more primitive. As years of accumulated debris are scoured away, so the world is cleansed of most of its human population. Those that are left desperately map the new landscape which encloses them in an attempt to make sense, and find patterns, in what they see. Yet as the heat and moisture of the fetid air encloses them, sense is difficult to find. They begin to dream or hallucinate. As they see visions of a blazing sun surmounting a strange waterlogged land, the protagonists come to realize that the languid nature of their own heartbeats has put them in touch with the past, not the future. Mankind is de-evolving, and inexorably, the entropic nature of man's own mortality is revealed in the text.

Compared many times over with Joseph Conrad's *Heart of Darkness*, Ballard's novel shares some of the former's characteristics. Insane hunters and native hordes populate a novel which at times seems to evoke a physical response in its reader with its oppressive atmosphere. The section in which the protagonist Kerans descends into the water-filled bowels of the London Planetarium lets Ballard remind us that we are dealing here with forces which eclipse the individual in dimension and time. 'The great dome of the planetarium hove out of the yellow light, reminding Kerans of some cosmic space vehicle marooned on earth for millions of years and only now revealed by the sea'[19]. As Kerans finds himself trapped in the main dome of the planetarium, gazing at the constellations before him, he slowly becomes oxygen deprived, and conscious of the enormity of scale, he has a vision: 'Epochs drifted. Giant waves, infinitely slow and enveloping, broke and fell across sunless beaches of the time-sea, washing him helplessly in its shallows. He drifted from one pool to another, in limbos of eternity, a thousand images of himself reflected in the inverted mirrors of the surface'[20].

Ballard's buoyant images are reminiscent of the imagery employed by Timothy Leary, the LSD guru, when he describes the floating associated with the higher levels of consciousness that he attributed to the sustained use of hallucinogenic drugs.

A similar approach was taken by Stanley Kubrick in his 1968 classic piece of science fiction cinema *2001: A Space Odyssey*. Written by Kubrick and Arthur C. Clarke, this complicated yet fulfilling piece of cinema explores the future of human evolution and technology, man's employment of artificial intelligence, and first contact with extraterrestrial life. It is distinguished by its attempts at scientific realism and its method of communicating emotions, thoughts and incomprehensible sights through psychedelic visual imagery. *2001* gained a substantial following amongst the New Age counterculture. It attempts to portray the inner mental landscapes of the protagonist Dave Bowman as an ornate artificial environment in the form of a Louis XVI-style hotel room. This artificiality stands in contrast with the flora and fauna which Ballard uses in *The Drowned World* to represent the inner landscapes of his hero's own mental architecture. Ballard's naturalistic approach was to strike a chord with many who read the text – perhaps because the New Age was a time when environmental issues came more sharply into focus.

Silent springs – Rachel Carson

It was during the New Age that the mass movements associated with environmentalism started to emerge in a coherent manner. This was in no small part due to the role of a single work – Rachel Carson's *Silent Spring*. Although it by no means initiated the environmental movement, it did serve as a focus to attract both support and criticism for its cause. The mass media response to

Carson's 1962 text was immense and served to open up new panoramas to communities hereto unaware of the potential problems associated with the developing consumer-led lifestyle of the New Age.

Carson, a writer, scientist, ecologist, and native of rural Pennsylvania, graduated in marine biology and zoology before working for the US Bureau of Fisheries as a science communicator. She also wrote radio scripts and contributed to the *Baltimore Sun* newspaper as a science journalist. Her career exploded when in 1936 she was appointed as Editor-in-Chief of all publications for the US Fish and Wildlife Service. An accomplished writer, in addition to numerous pamphlets on conservation and many scientific articles, she also authored five books. The first three of these focused on her acknowledged field of marine conservation and have been said by some commentators to constitute a 'biography of the ocean'. They were so successful so as to be translated into over thirty languages. Her books became bestsellers all over the world because they removed the rigid reserved language of science and substituted it with a more expressive style of writing. In the process they made Carson famous as a naturalist and science writer for the public. On the back of her success, Carson resigned from government service in 1952 to devote herself to her writing. It was during this period that she produced her magnum opus – *Silent Spring*. Initially appearing serialized in three parts in the June 1962 issues of *The New Yorker* magazine it was published in book form later that year and made a huge impact.

Given Carson's own exemplary record in both the fields of science communication and science writing it should come as no surprise that she chose to open the book with a science fiction short story. Penned by Carson herself and entitled *A Fable for Tomorrow*, this dystopic tale chronicles a small town in the turmoil of an environmental apocalypse when death falls from the skies in the form of a white granular powder 'like snow upon the roofs'[21]. As a consequence 'a strange blight swept over the area and everything began to change. Some evil spell had settled on the community: mysterious maladies swept the flocks of chickens; the cattle and sheep sickened and died. Everywhere was the shadow of death'[22]. The short story ends with the acknowledgement that the townsfolk had brought disaster on themselves, being responsible for their own destruction through the profligate use of agri-chemicals.

As an introduction, A Fable for Tomorrow arrests the reader, allowing the central disputation of Silent Spring. This assertion is that the reckless use of artificial chemical pesticides which had been saturating the agricultural industry throughout the Atomic Age could indeed lead such a nightmarish scenario. Both fiction and science combine to make a not so silent plea: that a more coherent and restrained approach to the employment of agri-chemicals would have substantial benefits. Silent Spring was a warning shot, fired across the bows of both the agricultural and chemical industries, as well as her former employers, the US Government – a genuine attempt to explain that they were in danger of underestimating the havoc that the long-term unchecked use of pesticides would have upon the environment. Carson's philosophy was inspired partly by the theologian Albert Schweitzer, to whom she dedicated the book. It was based on the notion of the interdependence of humanity and nature, and was illustrated through the text by the careful construction of scientific reasoning. Demonstrating her thesis, she contended that industrial activity causes lasting harm to the Earth's ecosystems. Focusing on the haphazard and sustained distribution of pesticides like DDT, she was able to demonstrate how toxins inserted into the food chain have severe, unpredictable, and far-reaching ecological consequences.

Testifying, just before her death from breast cancer, to the United States Congress in 1963, Rachel Carson demanded new policies to protect both human health and the environment. These demands resulted in DDT (the most notorious of the pesticides) being banned in the United States of America and throughout most of the world. Attacked by the chemical industry and parts of the government as an alarmist, Carson's approach of using A Fable for Tomorrow as an introduction to her subject opened her up to criticism. George C. Decker, an advisor to the US Department of Agriculture, commented that he regarded Silent Spring 'as science fiction, to be read in the same way that the TV program "Twilight Zone" is to be watched'[23]. Later commentators have also questioned the consequences of her actions: 'Carson didn't seem to take into account the vital role [DDT] played in controlling the transmission of malaria by killing the mosquitoes that carry the parasite.... It is the single most effective agent ever developed for saving human life... the anti-DDT campaign she inspired was responsible for almost as many deaths as some of the worst dictators of the last century'[24].

Despite some arguing that controversy was the ultimate legacy of Carson's work, the real inheritance of Silent Spring was a new awareness amongst people throughout the world that environmental vandalism was being perpetrated by the human race. Although science fiction had dealt with environmental themes in the past, most notably with M. P. Shiel's The Purple Cloud, in which during a race to North Pole the protagonist Adam Jeffson inadvertently escapes a worldwide disaster by which a gaseous purple cloud destroys almost all global fauna, leaving him as the sole survivor of a ravaged Earth. Despite the prose being at times as purple as the eponymous cloud, the scenes of isolation are convincingly done. The vandalism outlined by Carson was instead the concurrence of man's own calumny. It was this approach of flagrant disregard which came to be chronicled in the science fiction stories of the time.

Population explosion – John Brunner

New Age films tackled this topic head on, in films like the 1971 Soylent Green (based on the earlier Harry Harrison short story Make Room! Make Room!), depicting an Earth ravaged by years of environmental abuse and suffering from overpopulation and mass starvation, and the 1973 Silent Running, in which all that is left of the Earth's vegetation has been shipped into space in huge agri-domes. In these movies we have mankind's vandalism writ large across the big screen.

While not as popular as the films of the time, the works of authors like John Brunner whose New Age contributions included the seminal dystopias *Stand on Zanzibar* (1968) and *The Sheep Look Up* (1972) also address these themes. These satirical novels are firmly part of the British new wave. They mixed action up in the characteristically jump cut style so beloved by Moorcock, Ballard and the rest of the young Turks of the New Age. Brunner interweaves characters, locations and action seamlessly to bring us, in *Stand on Zanzibar*, a savage warning of the perils of overpopulation. *The Sheep Look Up*, in which an environmental holocaust which leads to the continental United States aflame, is seemingly culled straight from the pages of Carson's nightmares:

> When did you last bask in the sun, friends? When did you last dare drink from a creek? When did you last risk picking fruit and eating it straight from the tree? What were your doctor's bills last year? Which of you live in cities where you don't wear a filtermask? Which of you spent this year's vacation in the mountains because the sea is fringed with garbage?[25]

Paul Ehrlich takes Brunner's point seriously in his 1968 scientific frightener *The Population Bomb*. In this warning on the perils of overpopulation, mankind can no longer escape the onslaught of its own loins. Instead of the positivist view taken by Tsiolkovsky in the Astounding Age, in which mankind escapes its destiny through expansion into space, Ehrlich condemns mankind to fester in the ruins of its own disaster: 'The battle to feed all of humanity is over. In the 1970s and 1980s hundreds of millions of people will starve to death in spite of any crash programs embarked upon now. At this late date nothing can prevent a substantial increase in the world death rate'[26]. Ehrlich's solution to the problem is to adopt a form of Baconian eugenics, a series of population control measures exercised across the entire globe: 'Our position requires that we take immediate action at home and promote effective action worldwide. We must have population control at home, hopefully through changes in our value system, but by compulsion if voluntary methods fail'[27]. That Ehrlich's science was driven by the nightmare visions of fiction is surely not in dispute.

The perils of an overpopulated world were depicted first by such science fictional luminaries as James Blish and Norman L. Knight in A Torrent of Faces, with a trillion humans occupying just a thousand cities. Similarly, Isaac Asimov charted shining megacities, bursting at the seams with human life, in The Caves of Steel (1954). Frederick Pohl's The Space Merchants (1953), which has each step on a public staircase rented out as an individual apartment, is reflected in J. G. Ballard's 1962 short story Billennium in which future individuals find their own living space reduced to only three square metres. Brunner's was merely the latest voice to raise its concerns, which were able to be articulated in scientific circles through Ehrlich's tome.

The solutions advocated by both Ehrlich and a central protagonist at the end of Brunner's The Sheep Look Up were extreme – population control through involuntary statute and the mass suicide of entire continental United States:

> Don't keep the world on tenterhooks, Tom! Out with it! What's the best thing we can do to ensure a long, happy, healthy future for mankind? We can just restore the balance of the ecology, the biosphere, and so on – in other words we can live within our means instead of on an unrepayable overdraft, as we have been doing for the past half century – if we exterminate the two hundred million most extravagant and wasteful of our species[28].

Yet such extreme views served to fuel the debate and raise the issues high into the public consciousness. The moral panic created by the overpopulation rhetoric developed by Ehrlich and others in response to nightmarish visions conjured up by science fiction writers was to allow Nobel prize-winning scientists like Norman Borlog to pioneer agri-genetic methods of feeding the world. These approaches themselves raised the issues of mankind's dominion of the natural world around him and led future science fictional writers to explore genetic manipulation in a more sustained manner.

Humanity's control over its environment was a theme which was not just explored by scientists like Carson and authors like Brunner, but also by those who bridged the divide between the two, communicators and journalists like Frank Herbert. Although a published science fiction writer at the beginning of the New Age (with two pulp stories under his belt) Herbert still made nearly all of his money from his journalism. It was whilst researching a magazine article on coastal erosion and sand dunes in Oregon[29] that Herbert started to amass information on ecological matters. This research came to be the basis for his seminal work, *Dune*. Originally serialized as two short works in *Analog* magazine in 1963 and 1965, Herbert signposted its ecological content with the dedication which read 'to the people whose labors go beyond ideas into the realm of "real materials" – to the dry-land ecologists, wherever they may be, in whatever time they work, this effort at prediction is dedicated in humility and admiration'[30].

Set on the desert planet of Arrakis, *Dune* tells of the attempts of a young nobleman, Paul Atreides, to avoid the political machinations inherent in the feudal system operated through the galaxy whilst at the same time leading the indigenous Fremen to a new freedom as their messiah. The aboriginal Fremen, as depicted by Herbert, live closely entwined with their desert environment, battling water shortages, sandstorms and the massive native sandworms. During the course of the book it is revealed that their principal aim is the transformation of their inclement homeworld to a lush idyll. The Fremen are initially led by the native ecologist Dr Kynes, but Atreides later takes on the task of leading his people in this Herculean effort.

What is significant is that throughout the course of the narrative not a single character questions the wisdom of changing the climate of Arrakis. It is not until we find a brief description buried in an appendix do we realize that the ecologist Kynes and the hero Atriedes are seeking to restore Arrakis to its traditional nature: 'His thopter, flying between stations far out on the bled, was blown off course by a storm. When the storm passed, there was a pan – a giant oval depression some three hundred kilometres on the long axis – a glaring white surprise in the open desert. Kynes landed, tasted the pan's storm cleaned surface. Salt. Now, he was certain. There'd been open water on Arrakis – once'[31].

In the sequels to *Dune* we learn that in the Fremen, whilst successfully transforming their world into a verdant paradise, simultaneously undermine their societal system – a system predicated on the harsh realities of the arid planet. It is these complex questions on the nature of intervention into ecological matters which Herbert explores. Most significantly, *Dune* explores how justified is it to artificially intervene to stem the tide of ecological change, man-made or otherwise. Herbert's story extends Carson's hypothesis by asking whether such intervention should be made. Ultimately the lesson of the *Dune* series, learnt via the pseudo-religious teaching of the messianic Kwisatz Haderach of Paul Atreides, is that to fight the natural order of life, would have dire consequences. Herbert uses his texts to dismiss Descartian reductionism, arguing instead that whenever we try to direct nature, there are outcomes in many areas, some of which we will not understand until many years later. Herbert's holistic approach calls for the careful and structured planning of all environmental changes, a philosophy he put into practice when he later worked in Vietnam and Pakistan as both a social and an ecological consultant.

Sex, science and the secretary – star treks and wars

Some science fiction fans get very hot under the collar when discussing how influential Herbert's *Dune* was on the creation of the defining piece of twentieth century science fiction cinema, George Lucas's *Star Wars* (1977). Although Lucas has acknowledged that he had read *Dune*, the coincidental use of the desert planet of Tatooine, a planet-spanning galactic empire and imperial shock troops has led some fans to accuse Lucas of not so subtly adapting Herbert's text for the big screen. Yet *Star Wars'* obvious place as the cinematic incarnation of the space operas of the Astounding Age means that its philosophical significance and attempts at cultural exploration predate Herbert's own work. Any further resemblance to Herbert's texts needs to be couched in their shared common origins of the fantastic, the idea that both represent an incarnation of the monomyth, a shared cultural story couched in terms of the fantastic and explored in Joseph Campbell's *Hero with a Thousand Faces*.

George Lucas explicitly acknowledges Campbell's 1949 work as being very influential on the first draft of *Star Wars*. It charts the rise and rise of a young male adventurer who distinguishes himself in various fields of endeavour[32]. Collating the practice of cultures and societies around the world it summarizes the fantastic adventure story, of which science fiction is the latest adaptation, into a single monomyth. Herbert's text fulfils almost all of the Campbellian criteria for the monomyth, not least in the fact that Paul Atreides, the messianic figure in *Dune*, is (almost inevitably for its time) male. For whilst Herbert permeates his book with strong pivotal men as characters, women seem largely ignored. From the resourceful Lady Jessica and her Bene Gesserit sisters to the Fremen women who labour under the clannish rule of a male hierarchy, women come out of the text as confined by both class and tradition. The hierarchical system adopted by Herbert only seems to allow women success through marriage or by association. When they try to subvert the social system for their own ends the Bene Gesserit sisterhood are described on multiple occasions as 'witches'.

This inherent sexism saw its roots in the Astounding Age, with the pulps' emphasis on square-jawed heroes saving damsels in distress. This motif extended through the Atomic Age with female scientists being reduced to laboratory assistants or female superheroes (most notably *Wonder Woman* at the 'Justice Society of America') performing the function of secretaries. On the burgeoning small screen science fiction followed the same pattern. The televisual breakthrough of science fiction began with the BBC in Britain and their *Quatermass* series, which first aired in 1953. Incredibly popular, they had a reputation for emptying the pubs. The BBC followed this up with a specially commissioned new science fiction series which was to become one of the two most successful franchises in television science fiction. *Doctor Who*, which debuted in 1963, was remarkable in so many ways. Groundbreaking in concept and delivery, it still relegated the role of women to companions at best or the screaming damsels at worst, an approach which was to be followed by the other most successful science fiction franchise to emerge from television, Gene Roddenberry's *Star Trek*.

Clothing its female crew members in uniformed miniskirts, *Star Trek* relegated women to either 'the secretary', with either Lt Uhura or Yeoman Rand politely asking Captain Kirk if she can take a message, or the ubiquitous 'nurse', this time called Christine Chapel and played by Roddenberry's future wife, Majel Barrett. Barrett had to settle for this part after initially playing the first officer in the pilot episode, the explanation being that a woman with authority just wasn't believable. It was not until the very end of the New Age that television science fiction was to catch up with the growing representations of women in written science fiction. An excellent example with notable female characters most conspicuous is the 1978 British series *Blake's 7*, which has several strong female leads. The era of television science fiction was only beginning at this stage and although it did have a profound effect on the culture of its time (witness its depiction of the first interracial kiss – between Kirk and Uhura in *Star Trek* – to take place on American television) it is no surprise that it was lagging behind its written inspiration. The Computer Age was to see an explosion of television science fiction with much more innovation.

The second wave – Ursula K. Le Guin

Just as a new wave of writing was breaking on the shore of science fiction, so a second wave of feminism was starting to swell in the communities and societies of the Western world. Up until this point the focus of women's attention had rested upon *de jure* (officially mandated) inequalities, focusing specifically on universal suffrage and access to education. Second wave feminism merged *de jure* and *de facto* (unofficial) inequalities. By tackling issues like economic and reproductive rights, second wave feminism was resolute in adding social and economic equality to the almost full legal equality of many Western nations.

Beginning in 1963, the move towards equality was urged along by numerous women, two of the highest profile being Eleanor Roosevelt and Betty Friedan. Roosevelt chaired the Commission on the Status of Women, formed by the Kennedy administration, whose 1963 report acknowledged discrimination against women in virtually every area of American life. Friedan's *The Feminine Mystique* appeared on bookshelves later that same year – a collection of interview materials with women that bolstered the facts reported by the Commission. Like Zoline's *Heat Death*, its portrayal of a typical housewife trying to break free challenged assumptions:

> The problem lay buried, unspoken for many years in the minds of American women.... Each suburban housewife struggled with it alone. As she made the beds, shopped for groceries, matched slipcover material, ate peanut butter sandwiches with her children, chauffeured Cub Scouts and Brownies, lay beside her husband at night, she was afraid to ask even of herself the silent question: 'Is this all?'[33].

Once again science was to play its part; this time, the pharmacology of the New Age was turned towards reproductive rights and sexual liberation. Combining the chemicals of Leary with the research of Kinsey, the development of a viable oral contraceptive was seen as the key to women gaining both economic and sexual liberation. In particular, the control of reproduction, which the combination of the pill and access to abortion granted, allowed women to start to compete with men in a more significant way in all areas. It was these breakthroughs in science which were to provide the impetus for a whole generation of women science fiction writers. No longer encumbered by traditional notions of a woman's role in society, the reality of the new type of liberation afforded to women was to become extrapolated in the science fiction of the New Age:

> changes in the representation of women in science fiction have done little more than reflect the legal and social advances made by women in our society… which affect, for example, the position of women in the workplace or the demands made by women for greater autonomy in sexual choices[34].

Nowhere is this more clearly articulated than in the introduction to a masterpiece of science fiction, written at the time by a woman whose topics explored sexuality, gender and visions of power. In her own introduction to the later 1978 edition of her 1969 classic *The Left Hand of Darkness*, Ursula K. Le Guin reminds her readers that 'I write science fiction, and science fiction isn't about the future. I don't know any more about the future than you do, and very likely less'[35]. Le Guin's assertion is that she is writing to inform the contemporary debate using the medium of science fiction, an approach which is echoed throughout most of her work. Perhaps deriving from her background, the daughter of the cultural anthropologists Alfred and Theodora Kroeber, Le Guin followed her mother in using the medium of writing to explore anthropological themes through her science fiction.

Certainly *The Left Hand of Darkness* was hugely influential, winning Hugo, Locus and Nebula Awards. The plot focuses on the protagonist Genly Ai, himself, like Le Guin's parents, a cultural anthropologist who visits, as an ambassador, the alien planet of Winter in an attempt to get them to join the benign Ekumen federation of planets. As Ai flounders amidst the shifting political machinations of the varied factions of the planets he struggles to comprehend the inhabitants of the world, largely as a consequence of both their gender and sexuality. Predominantly gender neutral, the Winter planet dwellers are androgynous for 22 days out of 26 and then pick their gender depending on both company and need. The ambisexuality of the natives allows Le Guin to help the reader focus on the dual sexuality of our own societies. Relationships derived on power, masquerading as gender, become the focus for this important text. In Le Guin's own words:

> The question involved here is the question of the Other – the being who is different from yourself. This being can be different from you in its sex; or in its annual income; or in its way of speaking and dressing and doing things; or in the colour of its skin, or the number of its legs and heads. In other words, there is the sexual alien, and the social aliens, and the cultural aliens, and finally the racial alien[36].

A similar approach was explored by Marge Piercy in her *Woman on the Edge of Time* (1976), although this story was fused with a time travel element. This enables Piercy to explore more easily issues of gender difference in the contemporary politics of the New Age, an approach which bridges the divide between LeGuin and another great feminist writer of the New Age, Joanna Russ. Whereas Le Guin used the setting of an alien planet with alien mores and codes as a mirror with which to reflect our own societal differences and alienation, Russ hits much closer to home. Her great text *The Female Man* (1975) collects a protagonist, Joanne, from our own reality with three of her counterparts from

associated yet different realities. Using the fictional device of slid-
ing between these realities, we encounter first Janet, who lives on
the idyll of Whileaway, a version of our own reality in which men
have not existed for hundreds of years. The focus of the decentralized Whileaway society is to explore what it means to be
human whilst being unencumbered by notions of gender. We are
then introduced to Jeannine, whose reality is more akin to our
own while not having experienced the Second World War. Thus,
deprived of the technological and political impetus which that
conflict provided, Jeannine's world is stuck in a pseudo-Victorian
conservatism, a male-dominated order in which women didn't
work in the munitions factories and society wasn't denuded of a
generation of young men.

Finally, we meet Jael, a gender assassin who deals death to men
upon a world in which the revolutions associated with the gender
divide have taken the most extreme physical form: a never-
ending cycle of violence. Written in the cut-up style so closely
associated with the New Age, Russ's story flicks backwards and
forwards between the different realities. In doing so it allows the
reader to contrast not just the nature of each but also the interaction between their inhabitants. At times she removes her protagonists from the security of their own environment and immerses
them in the alternate, the alien, in the same way that she transports the reader hither and thither. Her central criticism of the
way in which women had been suppressed by society reflected the
findings of the Roosevelt Commission and Friedan's interviews:

Men succeed. Women get married. Men fail. Women get
married. Men enter monasteries. Women get married. Men
start wars. Women get married. Men stop them. Women get
married. Dull, dull[37].

Le Guin, Piercy and Russ all focus on the innerspaces of their
protagonists. Le Guin's Ai roves across the surface of the planet
Winter, persecuted and chased from one extreme to the other,

whilst the 'J's of Russ' *The Female Man* slip sideways between worlds, allowing the landscapes of their lives to remain mutable and carrying with them the permanence of their own inner architecture. These authors use their texts to explore and discover new realities by journeying to other worlds, a theme which was writ large across the science of the time. The advent of the space race placed within the consciousness both the practical measures taken to travel to the stars but also the realities that we would find there. As we gazed at our reflections in the surface of Voyager 2's golden discs who did we see? Ourselves or, beyond the tain, a different version of ourselves. The concept of different realities explored by Russ became in the New Age not just a metaphor for gender realities but a way of gauging the nature of those realities themselves.

The idea of alternate realities found expression first in the literature of the fantastic, from the mediæval concept of the faerieworld to Lewis Carroll imaginatively tumbling Alice down a rabbit hole. It only started to be explored with a scientific perspective when such concepts were played out by science fiction authors. It was H. G. Wells who arguably created the first explicit paratime novel with his 1923 work *Men Like Gods*, which features not only a multiverse theory but also a paratime machine. It was, however, with the New Age that it came to have any sort of scientific basis. Confined previously to time travel variants and historical exercises, in the New Age this type of storytelling was to mature. Nowhere is this maturity seen more clearly than in the work of Philip K. Dick, and in particular his 1962 work *The Man in the High Castle*.

Science was slow in catching up with fiction. Hugh Everett III had substantially mapped out a many worlds interpretation of quantum theory in his 1957 paper in *Reviews of Modern Physics*, Vol. 29, alongside a sympathetic review by his PhD supervisor John Wheeler. However, the many worlds interpretation was ignored for over a decade after its publication. It was left to science fiction mainstays of the New Age, like Michael Moorcock, J.

G. Ballard, Joanna Russ and Philip K. Dick, to bring the concept to the forefront of the public's attention. It was not until 1970, when Bryce DeWitt wrote a paper on Everett's work for *Physics Today*, that the scientific community started to pay attention. An anthology of work based on Everett's concepts was published in 1973, and the wheel turned full circle when a review of the work appeared in the science fiction magazine *Analog*.

At every step of the way science fiction had been willing to entertain concepts and principles which a conservative and hide-bound scientific establishment sidelined. When Everett met Niels Bohr, the father of quantum theory, in Copenhagen to discuss his work, Bohr cast it aside as inconsequential. Yet a concept that had already grasped the public imagination via science fiction could not be so easily dismissed. As Dick demonstrated and countless others followed, the idea of alternate realities was not going away.

Dick and Le Guin graduated in the same class from the same school, yet had the distinction of not knowing each other, as Le Guin had been moved up a year whilst Dick spent most of his final year absent from school with agoraphobia. Le Guin was to become one of Dick's greatest admirers, writing extensively about his work, calling attention to his approach and even going so far as writing her novel *The Lathe of Heaven* as an *homage* to him, featuring as it does Dickian themes of conflicting realities.

Grasshoppers and guns – Philip K. Dick

A troubled man, Dick spent most of his years battling with mental illness, a theme he was to visit in much of his work, along with a tendency towards pharmacological relief of both the legal and proscribed kinds. This, alongside the sequence of broken relationships which littered his life, gave his own reality a fractured feeling. His early works concentrated on sociological and political explorations of the other, moving only later to the alternatives which drugs and divinity could create. A true disciple of the New Age, Dick seemed to embrace with all his life the contradictions inherent in this schizophrenic time.

Prominent in Dick's work are the ideas of alternate universes and simulacra, as can be seen in the plot for *The Man in the High Castle*. Set in 1962 (the year the novel was published) it portrays a divided United States split between the victorious Axis powers of Nazi Germany and the Empire of Japan. Interweaving several storylines, Dick slowly allows the pattern of the novel to emerge. One plot focuses on espionage and attempts by the Japanese to discover a German plan to launch a nuclear attack on their home islands. Another explores the domestic lives of Americans operating under the government of the Axis powers. Central to these characters is one Frank Frink, who manufactures and distributes fake Colt pistols, as authentic artefacts of the American Civil War, to the Japanese occupiers, who are desperate for some real history. When one of his fake guns is used by a client to successfully defend himself from Nazi agents it calls into question the notion of what is real. Ray Calvin, another character involved in the fake memorabilia scam, makes this point to a girl he is seeing. He gives her two very similar-looking cigarette lighters, one of which is worth 'maybe forty or fifty thousand dollars on the collector's market'. Why? Because of the historicity:

> She said, 'what is historicity'? When a thing has history in it. Listen. One of those two Zippo lighters was in Franklin D. Roosevelt's pocket when he was assassinated. And one wasn't. One has historicity, a hell of a lot of it. As much as any object ever had. And one has nothing.... You can't tell which is which. There's no 'mystical plasmic presence', no 'aura' around it[38].

The notional of what is real and what is fake extends beyond artefacts to the realities in which they inhabit and also to the way that choices are made. Reflecting the changes in perspective is the pivotal section of *The Man in the High Castle*. In it, Frink encounters a bestselling underground novel entitled *The Grasshopper Lies Heavy* which comes to signify the existence of other potential realities, including one in which the allies won the war. The author of *Grasshopper*, Hawthorne Abendsen, reveals that, in common with many of the other characters in the book, he used the I Ching to guide his work. In a parallel to his own work Dick has revealed that he similarly used the divination process in writing the text. In doing so, Dick is explicitly acknowledging the influence that Taoist notions of balance (as personified in the coupled Yin and Yang) had on his writing[39], notions which he shares with his contemporary Le Guin, who similarly acknowledges[40] such influences in her creation of *The Left Hand of Darkness*. The innerspace became as important in the creation of the science fiction of the New Age as outer space.

A fusion of these two themes was explored in the 1977 cinematic blockbuster *Close Encounters of the Third Kind*. Directed by Stephen Spielberg, this tale of first contact has the various contactees being misled and misdirected by a government conspiracy to keep the landing site a secret. It is notable for its five-note musical contact sequence as well as the presence of the *nouvelle vague* director Francois Truffaut as one of the actors. *Close Encounters* takes the Dickian themes of the nature of reality and what we know to be true and plays them out across the backdrop of the Universe. When finally the hero Roy Neary makes contact with the mysterious 'greys', Sagan's final great demotion of mankind from the centre of the Universe comes to pass: mankind is no longer alone and a new reality is waiting to be explored.

Dick's volume of work is immense given the relatively young age at which he died. Spanning 44 books and hundreds of short stories, his work has been adapted for some of Hollywood's best known science fiction films. Steven Spielberg, for instance, moved on from *Close Encounters* to direct Dick's *Minority Report* in 2002. Yet Dick was only on the brink of worldwide success when he died aged 54, success that was going to be created largely through the big screen adaptation of a 1968 story, *Do Androids Dream of Electric Sheep?*, by the director Ridley Scott, as *Blade Runner* (1982). Undoubtedly influential in both its themes and cinematography, *Blade Runner* was to become a pivotal part of the new science fictional movement which was to encapsulate the scientific advances of the late twentieth century – Cyberpunk. With its fusion of gritty urban style and high-tech gadgets Cyberpunk commoditized everything it touched, proclaiming that the street had its own use for everything. The epic journeys of the New Age into both outer and innerspace were soon to be warped into visions of capitalism at its most feral.

References

1. Sagan, Carl, *et al.* (1977) *Murmurs of Earth: The Voyager Interstellar Record*. Random House, New York.
2. Sagan, Carl, *et al.* (1972) A message from Earth. *Science*, **175**, 881–5.
3. Gene Rodenberry (1966) The opening narration to the original *Star Trek* series.
4. Van Vogt, A. E. (1950) *The Voyage of the Space Beagle*. Simon & Schuster, New York.
5. Papke, M. E. (2006) A space of her own: Pamela Zoline's 'The Heat Death of the Universe'. In *Daughters of Earth: Feminist Science Fiction in the Twentieth Century* (ed. Justine Larbalestier), Wesleyan University Press.
6. Moorcock, Michael (1968) Introduction. In: *Best SF Stories from New Worlds 3*, Berkley Publishing Corporation, New York.
7. Rose, M. (1982) *Alien Encounters: Anatomy of Science Fiction*. Harvard University Press, Cambridge, MA, p. 31.
8. LeFanu, S. (1991) *Feminism and Science Fiction*. Indiana University Press, Bloomington, IN, p. 97.
9. Hewitt, E. (1994) Generic exhaustion and the heat death of science fiction. *Science Fiction Studies*, #64, **21**.
10. Leary, T. (1966) *Turn On, Tune In, Drop Out on ESP*. Audio recording. Original release: ESP 1027, USA.
11. Huxley, A. (1977) Foreword to *Brave New World*. Triad, London, p. 9.
12. Heinlein, R. A. (1981) *Expanded Universe, The New Worlds of Robert A. Heinlein*. Ace Books, New York.
13. Heinlein, R. A. (1961) *Stranger in a Strange Land*. Putnam, New York.
14. Bradbury, R. (1960) *Fahrenheit 451*. Voyager, London.

15. Landon, B. (2002) *Science Fiction after 1900*. Routledge, London, pp. 153–4.

16. Moorcock, M. (1963) Guest editorial in *New Worlds*, quoted James, Edward, *Science Fiction in the Twentieth Century*. Oxford University Press, Oxford, p. 168.

17. Ballard, J. G. (1985) Introduction to the French Edition of *Crash*. Vintage, London, p. 4.

18. Hewitt, E. (1994) Generic exhaustion and the heat death of science fiction. *Science Fiction Studies*, #64, **21**, 289–30.

19. Ballard, J. G. (1999) *The Drowned World*. Millennium, London, p. 101.

20. *Ibid.*, p. 110.

21. Carson, R. (1963) A fable for tomorrow. In: *Silent Spring*, Hamish Hamilton, London, p. 4.

22. *Ibid.*, p. 3.

23. Graham, F. (1970) *Since Silent Spring*. Hamish Hamilton, London, p. 39.

24. Taverne, D. (2005) The harm that pressure groups can do. In: *Panic Nation* (eds. S. Feldman and V. Marks). Blake Publishing, London.

25. Brunner, J. (1972) *The Sheep Look Up*. Ballantine, New York, p. 354.

26. Ehrlich, P. R. (1971) *The Population Bomb*. Ballantine, New York, p. xi.

27. *Ibid.*, pp. xi–xii.

28. Brunner, J. (1972) *The Sheep Look Up*. Ballantine, New York, p. 456.

29. Herbert, F. (1980) Dune genesis. *Omni* Magazine, July.

30. Herbert, F. (1963) Dedication to 'Dune World'. *Analog*.

31. Herbert, F. (1968) *Dune*. Hodder & Stoughton, London, p. 569.

32. Campbell, J. (1949) *The Hero With A Thousand Faces*. Princeton University Press.

33. Friedan, B. (1963) *The Feminine Mystique*. Victor Gollancz, London, p. 15.

34. Green, J. and Lefanu, S. (1985) *Despatches from the Frontiers of the Female Mind*. The Women's Press, London, p. 1.

35. Le Guin, U. K. (1980) *The Left Hand of Darkness*. HarperCollins, London.

36. Le Guin, U. K. (1975) American SF and the other. *Science Fiction Studies*, #7, **2**.

37. Russ, J. (1985) *The Female Man*. The Women's Press, London, p. 126.

38. Dick, P. K. (2001) *The Man in the High Castle*. Gollancz, London, pp. 65–6.
39. Warrick, P. (1980) The encounter of Taoism and Fascism in PKD *Man in the High Castle*. *Science Fiction Studies*, #21, **7**.
40. Bain, D. C. (1986) The Tao Te Ching as background to the novels of Ursula K. Le Guin. In *Modern Critical Views: Ursula K. Le Guin* (ed. Harold Bloom). Chelsea House, London.

Chapter 6

INFORMATION WANTS TO BE FREE: THE COMPUTER AGE

On 28 March 1979, a nuclear power plant operating in the Three Mile Island Nuclear Generating Station in Dauphin County, Pennsylvania, USA, suffered a partial core meltdown. There was reputedly a sole victim despite five days of anxious turmoil. That victim lived nowhere near the plant and did not have any contact with it.

In a classic example of life imitating art, the science fiction thriller *The China Syndrome* (1979), was released three days before the real-life incident. The film depicts a near meltdown, whilst a film crew led by Jane Fonda's character are at the fictional plant doing an exposé on nuclear power. This led Fonda to start campaigning against nuclear power. The pro-nuclear lobby wheeled out the renowned scientist Edward Teller to counter-lobby in their favour. The extra pressure resulted in the septuagenarian scientist suffering a heart attack, an event which he later blamed on Fonda in a two-page spread of the *Wall Street Journal*: 'You might say that I was the only one whose health was affected by that reactor near Harrisburg. No, that would be wrong. It was not the reactor. It was Jane Fonda. Reactors are not dangerous'[1].

The anxiety that seemed to convulse the Western world during both the Atomic Age and the New Age deepened, if anything, during the Computer Age. The longstanding faith that science and technology combined were a force which could improve the human condition was called into question. Programmes like the Strategic Defense Initiative (SDI) seemed designed to encourage the proliferation of dehumanizing technology, not rein it back.

Worldwide environmental pollution and the nuclear incidents at Chernobyl and Three Mile Island suggested that the hazards of technology outweighed its virtues.

The Three Mile Island versus China Syndrome incident reveals more than just rancour between the two high-profile antagonists. Behind Teller's advertisement in the *Wall Street Journal* was Dresser Industries, the multinational conglomerate whose defective valves contributed to the Three Mile Island accident[2]. This global corporation had a finger in many pies and expanded greatly as a result of the climate of reduced regulation and corporate expansion which the Reaganomics of the early 1980s amplified.

The climate of incorporation, with the economic power of large multinational companies wedded to an emphasis on a technological answer to every problem, led to the creation of a new kind of science fiction. This science fiction was so thoroughly rooted in free market capitalism and a technophilic love of gadgetry that it in turn came to influence a whole new worldview. From its emphasis on control – of individuals, societies and governments – it took *cyber*, meaning 'to steer' in Greek[3]. From its complete rejection of that control and authority, it borrowed the phrasing of a similarly nihilistic cultural movement of the late New Age – punk. Together they fuse as a portmanteau, their essential dichotomy giving a hint of the energy that resides within the movement. *Cyberpunk*, the science fictional incarnation of the aggressive techno-capitalism of the early eighties was born. First coined by Bruce Bethke in his short story of the same name in the November 1983 issue of *Amazing Science Fiction Stories*, in truth Cyberpunk was birthed two years before by a bespectacled technophobic American writer, resident in Canada – William Gibson.

Neuromancer – Cyberpunk and William Gibson

In May 1981 *Omni* magazine published a short story by a little known writer, William 'Bill' Gibson, entitled *Johnny Mnemonic*. A furious blend of techno-noir, its neon-lit streets, reflected in the mirrored shades of the characters who paced down them, entranced its readers. Gibson followed this up with a second short story utilizing the same setting, called *Burning Chrome*, that was introduced by Bruce Sterling to the Austin writers' workshop. Sterling and Gibson came to be the twin stars around which Cyberpunk orbited – Gibson for his influential output and Sterling for his relentless promotion of Cyberpunk as a movement in itself.

Cyberpunk needed a text, a more substantive locus around which it was to draw in consumers, other writers and critics. Gibson was happy to oblige, and in 1984 the definitive book was published: Gibson's *Neuromancer* hit the streets. Justifiably famous for being the first venue to employ the term cyberspace[4], Gibson used it to refer to the electronic matrix through which his protagonist insinuated himself. Such is the importance of this conceptual creation and the influence that it was to play on the development of computer science that Mike Davis comments:

> the opening section of *Neuromancer*, with its introduction of cyberspace, is the kind of revelation – of a possible but previously unimagined future – that occurs perhaps once a generation. Charles Babbage's and Ada Lovelace's anticipation of a programmable computer in the 1820s, Friedrich Engels' 1880s prophecy of a mechanized world war and H. G. Wells' prevision of the atomic bomb in 1900 are comparable examples[5].

Ironically crafted on an electric typewriter[6], *Neuromancer* followed the life of a burned out 'console jockey' called Case. Forcibly separated from 'the matrix' by disgruntled employers from whom he had stolen, the story chronicles the criminal Case's attempts to recapture his life online. Helped by a female bodyguard and razorgirl mercenary, Molly, and with an Artificial Intelligence (AI) called Wintermute pulling the strings, Case stumbles through the plot never really understanding what is going on.

Consensual hallucinations

Neuromancer is that rare combination of both style and substance, the fusion of the two demonstrated from its famous opening line 'The sky above the port was the color of a television, tuned to a dead channel'[7], to its description of cyberspace itself: 'A consensual hallucination experienced daily by billions of legitimate operators, in every nation, by children being taught mathematical concepts.... A graphic representation of data abstracted from the bank of every computer in the human system. Unthinkable complexity. Lines of light ranged in the nonspace of the mind, clusters and constellations of data. Like city lights receding'[8].

Case spends his waking moments moving through the nightmarish post-industrial landscape of a consumer-led world – a dreamlike postmodern fusion of pop-culture orientalism and western capitalist caricature. Yet Case never seems quite comfortable in this concrete and steel construction. It is only when he 'jacks in'[9], that he really feels alive.

Neuromancer explores how the burgeoning sciences of AI, virtual reality and genetic engineering could develop. Long before popular culture started to become concerned with the consequences of the unchecked growth of such science, William Gibson was asking questions. Could the Internet be addictive? Gibson's anti-hero gets withdrawal symptoms when he is forced to abandon the science which has come to dominate his life. Is there a way of legislating the development of future science and technology or should governments stay out of it?

Gibson's dystopic future paints a bleak picture of capitalist corporations overwhelming modern society to place profits before people. These corporate representations so vivid in his matrix, with huge multinationals exerting a global presence, find their ultimate expression in the form of the transnational family cartel of Tessier-Ashpool. It is one of the scions of this family who sought to create a future in which the family would use AIs to run the corporation and control cyberspace. The AI Wintermute, with Case and Molly's help, finally succeeds in putting this plan into effect. Fusing with the net, Wintermute gives it an intelligence all of its own only to reveal that he is not alone. He has detected evidence of another AI in a series of transmissions from the Centauri System[10]. The Computer Age traces the fall of humanism from the accusation that its cold, calculated reason ultimately ended in the Holocaust to modern techno-science: techno-science conceived as a monad – a self-contained entity oblivious to everything except its own interests. Nowhere is this more evident than in *Neuromancer*.

Neuromancer was the first in what came to be titled as the Sprawl trilogy; it was followed by *Count Zero* (1986) and *Mona Lisa Overdrive* (1988), in which cyberspace was the key concept. Gibson followed this trilogy with a second, the Bridge trilogy of *Virtual Light* (1993), *Idoru* (1996) and *All Tomorrow's Parties* (1999), this time concentrating on the race to control the beginnings of cyberspace technology and using as its pivotal concept the idea of the cyborg.

All of Gibson's Computer Age work, with a notable exception which we will soon discuss, directly reflects contemporary technology. His genius was to extrapolate from that concurrent technology in such an original manner so as to inspire real-world innovation. Yet that was not his intention

> ... I felt that I was trying to describe an unthinkable present and I actually feel that science fiction's best use today is the exploration of contemporary reality rather than any attempt to predict where we are going... The best thing you can do with science today is use it to explore the present. Earth is the alien planet now[11].

The exception to Gibson's contemporary extrapolations lies in *The Difference Engine* (1990), his alternate history novel, co-created with both Bruce Sterling and Benjamin Disraeli[12]. The first of Gibson's collaborators on *The Difference Engine*, Bruce Sterling, was a fellow Cyberpunk author; the second, Benjamin Disraeli, was a nineteenth century prime minister of Britain. Sterling's own Cyberpunk works deal with familiar techno-amped hardcore themes. Yet his contribution to the field goes beyond being an author. More than anyone else it was Bruce Sterling who created Cyberpunk. He took a loose association of writers and, in the public mind (although often not in the writer's own) welded together a coherent movement. His tool for this marketing strategy was *Mirrorshades – The Cyberpunk Anthology*. Its introduction is widely touted and disseminated as the defining document of Cyberpunk:

> Cyberpunk work is marked by its visionary intensity. Its writers prize the bizarre, the surreal, the formerly unthinkable. They are willing – eager, even – to take an idea and unflinchingly push it past the limits... they often use an unblinking, almost clinical objectivity. It is a coldly objective analysis; a technique borrowed from science, then put to literary use for classically punk shock value[13].

Gibson and Sterling together collaborated in thinking the unthinkable, and created an alternative vision of the past to reflect the reality of the present and define the possibilities of the future.

Cows and computers – Gibson and Sterling

The Difference Engine posits a world in which Charles Babbage, known variously as the man who invented the cowcatcher[14], which sweeps cattle out of the way of locomotives, an accomplished codebreaker[15] and 'the patron saint of the programmable computer'[16], transforms Victorian society with his inventions. Victorian London is translated into a proto cyber-city as the result of Babbage's successful production of his early computer,

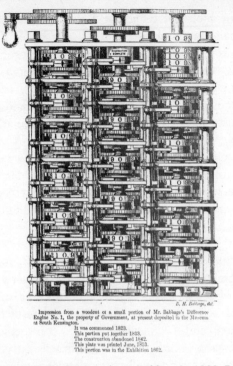

E. H. Babbage, del.

Impression from a woodcut of a small portion of Mr. Babbage's Difference Engine No. 1, the property of Government, at present deposited in the Museum at South Kensington.

It was commenced 1823.
This portion put together 1833.
The construction abandoned 1842.
This plate was printed June, 1853.
This portion was in the Exhibition 1862.

Figure 6.1 Original illustration from Babbage's 1822 Royal Astronomical Society paper 'Note on the application of machinery to the computation of very big mathematical tables'.

the Analytical Engine. Drawn from the collective consciousnesses of Gibson and Sterling, with substantial input from the long dead Disraeli[17], *The Difference Engine* dissects Disraeli's *Sybil* (1845) and uses its carcass to create a fresh narrative. Gibson and Sterling's counterfactual construction extrapolates the mass production of card-driven computers which revolutionize western society. These engines lead to a different outcome for the previously class-riven communities. The technological and informational revolutions of the late twentieth century are played out against a steampunk backdrop, as every man, woman and child is marked and counted by these vast machines:

the giant identical engines, clock-like constructions of intrinsically interlocking brass, big as railcars set on end, each on its foot-thick padded blocks. The whitewashed ceiling, thirty feet overhead, was alive with spinning pulley-belts, the lesser gears drawing power from tremendous spoked flywheels on socketed iron columns. White-coated clackers dwarfed by their machines, paced the spotless aisles[18].

This alternate history is our world conjured in the smog-stained streets of London. Great efforts are made to assert the similarities of the Computer Age with these cobbled streets. The problem is that Gibson and Sterling's efforts are 'not always smooth, and when Gibson and Sterling tweak over-hard to align "clackers" with "hackers", or to make London pavements into mean streets the *Neuromancer* audience will like, it can seem as if the past has been invested with arbitrary cool. Hey, these aren't dull Victorians. These are hip Victorians! Turbo Victorians! Victorians with attitude!'[19].

But behind the science fictional veneer both Gibson and Sterling are trying to make a serious point. They are pushing back the boundaries of cyberspace and the Computer Age. Instead of locating the matrix in the late twentieth century in totality, they are incarnating it earlier. For them the thrum of electric wires sending telegraph signals across continents was the first step. The telephone becomes a clear predecessor in its creation of a virtual communications net. For Gibson and Sterling the ubiquity of the computer finds its roots, its antecedents, in the pioneering science and engineering of the Victorian era. Information wants to be free – and it was people like Samuel Morse, Antonio Meucci and Alexander Graham Bell who first set about liberating it. As Sterling comments:

A science fiction writer coined the useful term 'cyberspace' in 1982, but the territory in question, the electronic frontier, is about a hundred and thirty years old. Cyberspace is the 'place' where a telephone conversation appears to occur. Not inside your actual phone, the plastic device on your desk. Not inside the other person's phone, in some other city. THE PLACE BETWEEN the phones. The indefinite place OUT THERE, where the two of you, two human beings, actually meet and communicate[20].

Science had created an environment in which authors like Gibson and Sterling wanted to play. Set free from the straitjackets of convention they set about creating their own consensual visions of both the future and the past. These visions were, in turn, to influence the next generation of scientists. Increasingly the Cyberpunk vision of the future, glimpsed briefly through a pair of mirrorshades, was being used as a blueprint for what was to come. The idea that everyone on the planet would interact with a computer and that most people would own one was shifting from a science fictional concept to the reality of the world at the end of the second millennium.

Ubiquity of computing

A Western Union memo of 1876 remarked that, 'This "telephone" has too many shortcomings to be seriously considered as a means of communication. The device is inherently of no value to us'[21]. Thomas Watson, chairman of IBM, commented in 1943: 'I think there is maybe a world market for maybe five computers'[22]. In 1977, Ken Olson, president, chairman and founder of Digital Equipment Corporation commented that, 'There is no reason anyone would want a computer in their home'[23], yet by 1981 Bill Gates takes for granted the future ubiquity of the Personal Computer (PC) with his alleged assertion that, '640K ought to be enough for anybody'[24].

In 1981 IBM stunned the world when on 12 August 1981, its first PC went on sale. Costing $3000 and with only 64 kbyte of RAM, a floppy disk drive and monochrome graphics, it seemed an unlikely candidate for instigator of a worldwide cybertech revolution. Although primarily focused on the business market, users around the world could see its potential. Within twelve months games were being launched, the first being a flight simulator, for the new machine. Yet it was not the PC which brought computing to the masses. That honour belongs to the Commodore 64, released only one year later.

The Commodore 64 was designed to be interfaced with the most influential consumer electronic product of the twentieth century – the television. As such it reached out and created an altogether new level of popularity, fuelled partly by its price of $400, partly by its availability in high street stores, but mostly by the presence of games you could see in colour and hear via audio synthesizer chip. This was the product which brought the Cyberpunk vision of ubiquitous computing to life. It is the best-selling single personal computer model of all time[25].

Consumers in 1982 watched computers brought to life in the big-budget SF film *Tron*. What they saw on screen they wanted to participate in. The new generation of personal computing offered them that opportunity. *Tron* features a computer games programmer, played by Jeff Bridges, who gets digitized into a virtual computer-based world by a crazy computer control program AI known as the MCP (Master Control Program). Forced by the MCP's forces of law and order to fight in a series of gladiatorial games which echoed the products that Commodore users could purchase for their home machine, Bridges' character Flynn is aided by an incarnation of an icebreaker programme, the eponymous Tron. Ultimately defeating the technological tyrant, Bridges' character overthrows the system and all it represents so as to be reincarnated back into the flesh of the corporeal world.

Tron explores the limits of automation and computer control. The hero is imprisoned and forced to fight for his life. The computer, in the form of the MCP, is not depicted as a morally neutral piece of technology. Instead, *Tron* explores whether science can ever be neutral when it has been designed by fallible humans. At the end of the film Flynn emerges from the computer to take control of the software mega-corporation ENCOM which runs the MCP system – leaving the question hanging of whether ENCOM will be any better than under its previous operators.

One of the first films to use long CGI sequences, this Disney-produced special effects-driven extravaganza integrated about half an hour of computer-generated effects. Although CGI had been used first in the science fiction films *Westworld* (1973) and its sequel *Futureworld* (1976), it was in *Tron* that it found its natural home – which was a good thing, considering that the film is generally lambasted for an incoherent plot and uninspired acting. Ironically, whilst the film earned Academy Award nominations for costume design and sound, it was snubbed in the visual effects category. Competing against *E.T.*, *Blade Runner* and *Poltergeist*, Spielberg's diminutive extraterrestrial won because Academy voters reportedly felt that the filmmakers had cheated by using computers.

Consumers too felt cheated by computers. Many expected to be able to extract their new purchase from its box and become proficient hackers and cyberwarriors in an instant. The truth was that early personal computers were large machines that frequently crashed and were not easy to interface with. But the interface became user-friendly with the launch of the Macintosh by Apple. The Macintosh featured easier to use programs, 3.5" disks which were sturdier than their 5.25" PC counterparts and a superior WYSIWYG (What You See Is What You Get) word processing package. These innovations changed the way that people saw, and used, computers. Equally famous for its method of introduction as its advanced usability, it was introduced to a science fictional world which took seriously William Gibson's contention that 'the best thing you can do with science today is use it to explore the present. Earth is the alien planet now'[26].

Surprisingly for companies associated with groundbreaking technological innovation, the pioneering computer manufacturers had tried hard to steer clear of science fiction in the promotion of their machines. They were afraid of intimidating their audience: 'The problem was, when Americans thought of computers and science fiction, they imagined HAL of *2001: A Space Odyssey*. Companies wanted to reassure consumers that computers were simple, unthreatening devices'[27]. Indeed, IBM's first promotional material featured Charlie Chaplin.

Orwell's Apple

All that changed in 1984. Apple debuted the Macintosh via a television advertisement aired on 22 January 1984 during a commercial break in the third quarter of Super Bowl XVIII. Fusing together numerous science fictional icons, it was the work of Ridley Scott, director of *Alien* (1979) and *Blade Runner* (1982). The ad featured a blonde unnamed heroine who, like Dorothy in *The Wizard of Oz*, appears in colour in a monochrome world, with the same shades of the contrast between the neon and grey as Scott's *Blade Runner*. Dressed as an unencumbered athlete, including a top emblazoned with the apple logo, this iconoclast sprints through legions of bald stooped wage slaves drawn directly from Lang's *Metropolis*. Sarah Stein points out that 'The running woman of the Macintosh ad is reminiscent of Sarah Connor, the female protagonist of *The Terminator* released the same year – also a woman whose physique was womanly rather than masculine, and in whom sexuality was configured as a life force she could call upon in her warrior role'[28]. Darting through a dystopian Orwellian world she abruptly stops and hurls a sledgehammer at a TV image of Big Brother – an implied representation of IBM[29], whose voice is busy pronouncing in stentorian overtones:

Today, we celebrate the first glorious anniversary of the Information Purification Directives. We have created, for the first time in all history, a garden of pure ideology. Where each worker may bloom, secure from the pests purveying contradictory thoughts. Our Unification of Thought is more powerful a weapon than any fleet or army on earth. We

are one people. With one will, one resolve, one cause. Our
enemies shall talk themselves to death and we will bury
them with their own confusion. We shall prevail![30]

The final image of the advertisement fuses an identical visual
and auditory message; 'On January 24th, Apple Computer will
introduce Macintosh. And you'll see why 1984 won't be like
1984'[31]. The advertisement referenced Orwell's famous science
fictional dystopian vision of a world in which individuality is suppressed. Apple's campaign fused together the Cyberpunk hacker
values of subversion and revolution; 'PCs were introduced in the
1970s as tools – utilitarian objects... in the 1980s, they became
full-fledged commodities – shiny consumer products defined not
just by their use value, but by the collection of meanings, hopes,
and ideals attached to them through advertising, promotion, and
cultural circulation. With the 1984 ad, Apple identified the
Macintosh with an ideology of 'empowerment' – a vision of
the PC as a tool for combatting conformity and asserting individuality'[32].

Apple set itself up as the voice of the individual in a world of
dominant corporations (despite its own multinational, multi-
million corporate presence). In doing so it fed into the consensus
capitalist competition of computing, a competition which en-
shrined within it a new techno-scientific paradigm – Moore's
Law.

Moore's Law

At the heart of the scientific method lies empirical observation,
the observation of evidence and consequences. Such an empirical observation was made by Gordon E. Moore, a co-founder of
the electronics and computer giant Intel. Writing in *Electronics*
magazine in 1965, Moore argued that the number of transistors
per square inch on integrated circuits had doubled every year
since the integrated circuit was invented and that this trend
would continue for the foreseeable future[33]. This exponential

curve in data density has largely held steady since then. Such was the influence of Moore's prediction that it was accorded the title Moore's Law early in the New Age. By the beginning of the Computer Age it was used for just about anything in computing, its ubiquity a parallel to the spread of computing itself. Network capacity and operating speed became subject to this prophetic utterance. Moore's Law came to define the efforts of the entire era. In a triumph of self-delusion over scientific extrapolation, the entire industry came to regard it as the minimum benchmark. Thus it was transformed into a piece of self-fulfilling divination. Yet the exponential era proved inspirational for science fiction as Moore's Law led Cyberpunk authors and commentators like Ray Kurzweil, Vernor Vinge and Bruce Sterling to hypothesize about a technological singularity.

The singularity is the technological creation of smarter-than-human intelligence. Although commonly seen as referring to artificial intelligence alone, there are several other technologies that could lead to the singularity, including direct brain–computer interfaces, biological augmentation of the brain and genetic engineering. Leading singularity author and computer scientist Ray Kurzweil is known not just for his work in this area, but also for the creation of an artificial 25-year-old female rock star alter ego, Ramona, an artiste who frequently performs concerts through virtual reality technology. This development inspired the science fiction film *S1m0ne*, starring Al Pacino. The plot features a film director who creates a synthetic superstar who comes to dominate the movie industry. Seeking to mislead the world, the director is forced to spin further and further out of control as his illegal activities deepen.

Kurzweil projected an extension of Moore's law to encompass an exponential trend in the world's technological development so that the emergence of a smarter-than-human intelligence (AI) occurs in the mid-twenty-first century[34]. The key for Kurzweil is that he regards Moore's Law as only a small part of a wider principle, a theory he refers to as the Law of Accelerating Returns. This rule states that whenever a technology draws near to some kind of an obstruction, a new technology will be innovated to bridge the gap. The principle predicts that the frequency of such paradigm shifts will increase over time, so that this gradual process will amplify in tempo in what is referred to as a soft take-off:

> An analysis of the history of technology shows that techno-
> logical change is exponential, contrary to the common-
> sense 'intuitive linear' view. So we won't experience 100
> years of progress in the twenty first century – it will be more
> like 20,000 years of progress (at today's rate)... within a few
> decades, machine intelligence will surpass human intelli-
> gence, leading to The Singularity – technological change so
> rapid and profound it represents a rupture in the fabric of
> human history[35].

Kurzweil's idea of paradigm shifts creating a change in future societies finds a mirror in science fiction, which Amit Goswani reminds us is 'that class of fiction which contains the currents of change in science and society. It concerns itself with the critique, extension, revision, and conspiracy of revolution, all directed against static scientific paradigms. Its goal is to prompt a paradigm shift to a new view that will be more responsive and true to nature'[36].

Singular states – Vernor Vinge

Science fiction again drove science forward with regard to its conception of the singularity. Although Kurzweil and others like Marvin Minsky, co-founder of MIT's AI laboratory, and Hans Moravec, the 'father of AI', started to take the extrapolations of this concept seriously in the late 1990s, it was the Cyberpunk author and professor of mathematics Vernor Vinge who in 1981 heralded the whole process. Although Vinge's predictions do differ from Kurzweil's in that he predicts a sudden ascent to the singularity, rather than a soft takeoff, both embrace the same concept. Vinge confesses:

When I was first writing science fiction in the early '60s, it was easy to have ideas that it turned out didn't happen for 20 or 30 years... but now it is very hard to keep up – in part because the people who are making things happen have absorbed science fiction's mind-set of scenario building and technological brainstorming. They are driving ahead of their headlights too, now, and things are going very, very fast[37].

Vinge introduced the idea of the Singularity in his 1981 novella *True Names*, which, as well as innovating around this concept, also presented a detailed rendering of cyberspace, giving it a central position within Cyberpunk. In *True Names* a crowd of computer geeks living in a society on the cusp of the singularity are using a new type of fully immersive virtual reality technology referred to as the 'Other Plane'. Together they enjoy using it to participate in communal sword-and-sorcery type adventures. Vinge is by this anticipating the online computer games of the twenty-first century by some two decades. These illegal activities place them at odds with the forces of law and order in this world. Their greatest delight is wreaking mayhem and havoc in anything from petty vandalism to massive disruption. Whilst their avatars are online accomplishing their amazing deeds their recumbent bodies lies vulnerable and comatose in secret locations. But it all starts to disintegrate when a new avatar turns up and agitates for change. The activist is revealed as a neglected government AI bent on revenge. It starts to recruit other conspirators in a scheme in which the domination of cyberspace will directly parallel a similar domination of the outside world. Attracting unwanted attention, all avatars are forced to hide their true names or face real death when the authorities take an active interest in deleting them both in and out of the net:

they were experiencing what no human had ever known before, a sensory bandwidth thousands of times normal... a

tidal wave of detail rammed through the tiny aperture of
their minds. The pain increased, and Mr. Slippery panicked.
This could be the True Death, some kind of sensory
burnout[38].

True Names is a classic example of science fiction confronting
the inhuman. Androids, cyborgs and AIs as motifs of techno-science encroach ever further into our daily existence until the
spectre is raised of advanced artificial life-forms. From *2001*'s
HAL, to the Architect in *The Matrix* and the Mailman in *True
Names*, these artificial life forms are shown to be contesting our
domination of the planet and its resources.

Waking the web – ghost in the shell

The creation of an artificial life form such as the AIs envisaged as
a result of the singularity are part of a long process. As the award-
winning chemist Eamonn Healy remarks in Richard Linklater's
decorated animated film *Waking Life*, history has so far consisted
of:

> two billion years for life, six million years for the hominid, a
> hundred-thousand years for mankind as we know it... you're
> beginning to see the telescopic nature of the evolutionary
> paradigm. And then when you get to ten thousand years for
> agriculture, four hundred years for the scientific revolution,
> and one hundred fifty years for the industrial revolution...
> the new evolution stems from information[39].

Incarnating the idea that the next evolution is an informational one and fusing it to the rallying cry of the Computer Age –
information wants to be free – is the Japanese anime *Ghost in the
Shell* (1995) based on the earlier 1989 manga of the same title.
This animated classic focuses on the emergence of an AI, referred
to variously as 'Project 2501' and 'The Puppetmaster', from the
web itself. A superior animated movie in the trademark Japanese

style, it became very influential on the science fiction of the Computer Age. In a world where 'Humanity has underestimated the consequences of computerization'[40], and life has become dominated by corporate monopolies, everything is portrayed as reliant on a vast computer network. The Puppetmaster has been hacking into people's brains in an effort to find a permanent repository, eventually fusing with the film's cyborg heroine, Major Motoko Kusanagi. *Ghost in the Shell* has some serious philosophical extrapolations to make.

It looks at the notion of cyborgization. How much of our body can we replace and still be called human? It asks whether artificial life forms have souls, and if they have the right to reproduce. *Ghost in the Shell* explores all these amidst a fusillade of gunfire. It also explored the possible future of the Internet, and in its pervasiveness seemed to get it right.

Today the Internet is seen as omnipresent: a vast array of computer networks linked together and encompassing the globe. The truth is more humdrum; groups of computers are linked, via a variety of often mundane means, to exchange information. The Internet had its origins in the US government's determination to recapture the technological lead from the USSR following the launch of Sputnik at the end of the Atomic Age[41].

It was contemplation of the Internet's military origins which led science fiction to consider the consequences of the new development. The 1983 film *WarGames*, which sees a young hacker almost wreak thermonuclear war upon the Earth through a bungled attempt to impress a girl, has clear antecedents: 'clearly the plot of *WarGames* revives a set of anxieties previously articulated in the 60s, with films such as *Colossus – The Forbin Project* and *Dr Strangelove*, but this time with the added frisson of contemplating the potential insecurity of extended communications networks'[42].

On 29 October 1969 the first node of the Internet (then known as ARPANET) went 'live' at UCLA to be followed soon by the British Post Office, Western Union International and

Tymnet. Together they integrated to form the first international packet switched network, referred to as the International Packet Switched Service (IPSS), in 1978.

But it was not until the 1990s that the Internet gained a public persona. On 6 August 1991 the world's largest particle physics laboratory, which straddles the border between France and Switzerland, revealed the results of a two-year project helmed by Tim Berners-Lee known as the World Wide Web. The *Organisation Européenne pour la Recherche Nucléaire* (commonly known as CERN) project to integrate information effectively was revealed, appropriately, on the alt.hypertext newsgroup, heralding in one stroke the results of the new research, the application of the data and a new science fictional future.

An ephemeral informational confluence, the web has taken on a life of its own. Broken links and moribund pages slowly fester and die as they are removed by service providers. New content is being continually added as the web constantly re-evolves and reinvents itself for each new moment. Is the whole more than the sum of its parts? Is the web in some sense alive? Perhaps it is a new form of organism. It reproduces, and it has defences and an immune system with which its fights the diseases and bacteria of hackers and viruses. It also perpetuates itself by codifying information. Automated webcrawlers sent scurrying from Google and its cohorts constantly map and store information, trying to spin a picture of the web. If the web is artificial life of a type not seen before, then what kind of life would it be? What would its offspring look like?

Artificial life – Terminator

One attempt by science fiction to extrapolate answers came in the form of the hugely successful Computer Age science fiction *Terminator* franchise. Fusing the apocalyptic anxiety of *WarGames* with the idea of an emerging AI consciousness, the adversarial Skynet seeks to eradicate human life from the planet by deploying a wide variety of its chromed children in murderous mechanical mayhem.

The original 1984 incarnation of the franchise depicts a 'Terminator' android (played by Arnold Schwarzenegger) being transported back in time from 2029 to 12 May 1984 to assassinate a woman named Sarah Connor. Connor (played by Linda Hamilton) is the mother of the future leader of the resistance, John Connor. In an effort to stop history being altered the resistance send back a fighter of their own, who in the course of the movie manages to both save Sarah and impregnate her with her soon-to-be hero son. The success of the original movie took everyone by surprise, including the director James Cameron, who had commented 'we know that we're going to get stomped by the Christmas movies. *Dune, 2010,...* I'll be lining up to see them – why shouldn't everybody else?'[43]. Acclaimed not just by the general public but by critics and academics, the film was read as an anti-establishment, anti-capitalistic allegory. Its provocative message was that those in authority failed to heed the warnings of impending doom and that the development of Skynet by the Cyberdyne Corporation was a triumph for unfettered greed. It was augmented by a gun control agenda which led politicians to point out the accuracy of Schwarzenegger being able to walk into a gun shop and emerge as an armed and armoured Arnold[44]. Both Connor and her saviour Reese have little time for the police in this scenario, instead choosing to place themselves outside of the law; to do what the law cannot, they become that most attractive of archetypes – the anti-hero.

The success of the film led to calls for a sequel, not least from a public who seemed to identify more with Schwarzenegger's transtemporal assassin than with the wholly human protagonists. Thus in 1991 the sequel *Terminator 2: Judgement Day* was released. Reuniting Cameron as director with stars Hamilton and Schwarzenegger, Hamilton's opening voiceover set out the film's stall:

3 billion human lives ended on August 29th, 1997. The survivors of the nuclear fire called the war Judgment Day. They

lived only to face a new nightmare: the war against the machines. The computer which controlled the machines, Skynet, sent two Terminators back through time. Their mission: to destroy the leader of the human resistance, John Connor, my son. The first Terminator was programmed to strike at me in the year 1984, before John was born. It failed. The second was set to strike at John himself when he was still a child. As before, the resistance was able to send a lone warrior, a protector for John. It was just a question of which one of them would reach him first[45].

Judgement Day took the audience identification of Arnold Schwarzenegger and made a virtue of it by this time casting him as the hero, his now obsolete T800 model 101 series android being forced to duel with the fluid mimetic metal of the androgynous T1000. Whereas the original android was 'a cybernetic organism. Living tissue over a metal endoskeleton'[46], the new T1000 was made from a polyalloy capable of taking on any form.

It is not the transformation of the T800 from villain to hero that is the most striking. This comes in the person of Sarah Connor herself. She is changed from the caring sympathetic mother-to-be in the first film into a calculating relentless killer by her experiences. The well-muscled Connor will stop at nothing to protect her son, whereas the T800 is compelled by the young John Connor to take only non-lethal action. Moving from anti-hero to vigilante she chooses to take matters into her own hands, only to be pulled back from the brink by her more compassionate child.

The final film in the trilogy, *Terminator 3: Rise of the Machines* (2001), similarly re-united the T800 101 model with a previous companion, the now fully grown John Connor, living off the grid and making his way as a petty criminal. This time the antagonist was a gynoid, a seductive female Terminator – the T-X, referred to throughout the film as the Terminatrix. Christina Lokken's portrayal of the T-X fused not just the feminine with the mechanical, but a leather clad dominatrix with the Terminator. From the overtly masculine T800 via the androgynous T1000 we finally arrive at the feminine T-X. The T-X hybridized the core concepts of both the 800 and 1000 series models with both an endo-skeleton and poly-mimetic capabilities. This is by far the most capable adversary of the series, not only for its physicality but also its psychology, at one point inflating its breasts when it realizes that sexuality is as much a weapon as brute force. It presents the culmination of the development of the terminator artificial life form. Moving beyond the android origins of the T800 and inte-grating the amorphous fluidity of the T1000, the T-X becomes not just a gynoid but a cyborg: 'Contemporary science fiction is full of cyborgs – creatures simultaneously animal and machine, who populate worlds ambiguously natural and crafted. Modern medicine is also full of cyborgs, of couplings between organism and machine, each conceived as coded devices, in an intimacy and with a power that was not generated in the history of sexual-ity'[47].

The *Terminator* series becomes an exploration of what it means to be alive, to what lengths we will go to survive. John Connor, the result of a biological coupling, spent all of his life – *in utero*, as a child and as a man – fighting to avert an apocalyptic future. Yet the end of *Terminator 3* reveals that events conspired at every step to instigate that future. John's actions at the end of the trilogy look as much the results of a programmed existence as those of the various *Terminators* – those results of a coupling between organism and machine.

At stake in these Computer Age parables is the very belief of what it is to be human. Both *Ghost in the Shell* and *Terminator* alike question what if what is 'proper' to humankind is to be inhabited by the inhuman? Can we be programmed? Are we less valued if key parts of our bodies are not 'natural' tissue? How many syn-thetic body parts can we tolerate without losing identity? The ultimate extrapolation of these films becomes an exploration of how consciousness will be affected through augmentation.

Cyborg manifesto – Donna Haraway

The idea of the cyborg, the man–machine fusion, had been around in science fiction for a substantial amount of time. It was not until the Computer Age that, in both imagination and reality, the cyborg came to be realized in any momentous way. Although myth and legend from around the world told stories of created beings, fusions of metal and flesh, it was not until Shelley's *Frankenstein* that the Mechanical Age took seriously the idea of union between science and imagination to create new life.

The first serious scientific proposal of cyborgs was an offshoot of the Astounding Age. When speculating on the modifications needed for long-term space travel, the great British scientist J. D. Bernal, wrote[48] in 1929 that it would only be through adaptation that the prolonged exploration of space could be achieved. In this he included both prosthetic surgery and a hard-wired man–machine interface. By the beginning of the New Age this idea had been extended. Following a similar extrapolation to that of Bernal, Manfred Clynes and Nathan Kline, who worked at the US Rockland State Hospital's Research Laboratory, proposed at a NASA conference that the successful exploration of space depended on a vision of human and the spacecraft as interpenetrated systems which shared information and energy. It was Clynes who coined the term cyborg to reflect their positive idea of future developments which left 'man free to create, to explore, to think, and to feel'[49].

The popular visual fictions of the early Computer Age took the concept of the cyborg to their biomechanical heart. *Tron*, *Blade Runner*, *Videodrome*, *Akira* and *The Terminator* all shared a fascination with the man-machine which echoed the literary Cyberpunk fascination with the same. This was to have a profound effect on the field of consciousness studies. In 1985 the Computer Age found its cyborg seer. The US academic Donna Haraway published *A Manifesto for Cyborgs: Science, Technology, and Socialist Feminism in the 1980s*. Noted for her previous explorations of the preconceived notions of masculinity in science, Haraway's work burst into the Computer Age with unrestrained vigour. Haraway's *Manifesto* censures the traditional feminist emphasis on identity rather than affinity. Introducing the concept of the cyborg, Haraway wanted to move away from dualistic gender politics and instead embrace a world without limitations – a science fictional world. The contemporary feminist analysis of Cyberpunk was that, despite the attempts by its champion Bruce Sterling to create a genuine sense of gender sensibility, within the Cyberpunk movement much of its culture either stereotyped women in traditional roles or forced them to behave like men. The raygun Reagan polarized gender politics of contemporary society led to the feminist criticism that Cyberpunk was 'an alternative, attractive, but hallucinatory world which allows not only a reassertion of male mastery but a virtual celebration of a kind of primal masculinity'[50].

It also led to claims that the Cyberpunk movement was over-dominated by male authors, a claim which Pat Cadigan, the 'Queen of Cyberpunk', addresses. Cadigan is a notable Cyberpunk contributor whose short stories and novels directly contributed to the development of the movement. In her introduction to *The Ultimate Cyberpunk* (2002) she comments that 'Cyberpunk was never concerned with the biology of the writers involved, regardless of what anyone might think. To force the issue of how many men vs. how many women there are is simply another way to begin from an improper assumption'[51].

Haraway recasts the debate by arguing that a cyborg's built-in androgyny should be seen as a potential victory for feminism. For Haraway a cyborg is not only a fictional hybrid of machine and organism, a cybernetic organism, but also a creature of lived social reality[52]. The androgyny of the cyborg has the potential to drive considerable change within society. Hence the nature of Haraway's remarks being couched as a manifesto, a political statement of beliefs implicitly codifying a call for change. Haraway's work recognized an unspoken truth about the Computer Age. We had already become cyborgs.

The Computer Age saw the erosion of the man–machine divide. We came to understand ourselves, at least implicitly as cyborgs. Sitting watching science fictional visions of hybridity play out across the silver screen we adjusted our hearing aids, reset our glasses on our noses and even, when an exciting scene required, allowed our pacemaker to facilitate the transfer of blood around our body. We have been implanted with the very consumer goods we are manufacturing. As cyborgs we are wearing and integrating a consumer society into ourselves. We have become the products of the Computer Age. Our biomedical reality is one in which more and more of us are augmented, not just with glasses but with polarized Rayban Aviators – the mirrorshades of Cyberpunk. What was once medical necessity had become sublimated to consumer desire. But where will it end?

The biomedical reality of a world in which many feature artificial additions had started in the New Age, with the Thalidomide tragedy of the early 1960s alongside an ongoing Vietnamese conflict which had spawned numerous amputees amongst the American armed forces. This created an atmosphere in which prosthetic research flourished. This research was spearheaded by those working with the children affected by Thalidomide and within veteran rehabilitation and resulted in the successful development of new kinds of bionic prostheses. No longer was the goal just the replacement of the body part by a mechanical version. The target now was to either mimic the original function of the body part very closely, or even surpass it. Medical bionics were described in *Cyborg*, a novel by Martin Caidin. This is the story of a test pilot, Steve Austin, who undergoes extensive surgery after a catastrophic plane crash. The book details Austin's operation and his reaction both to his original injuries, including his attempts to commit suicide, and to his being rebuilt with bionics. When remade into a popular science fiction series, *The Six Million Dollar Man*, the emphasis of the book was changed to assure the viewer that Austin becomes 'better, stronger, faster'[53].

Extrapolating from the positivism of the *The Six Million Dollar Man*, the cyborgs of the Computer Age were given to using advanced prostheses to replace healthy body parts with artificial mechanisms and systems to improve function. No longer was prosthetics the notion of artificial replacement for loss of utility. Instead it became the augmentation and extension of function to surpass what had been replaced. Technology had lived up to the science fictional dictum of Cyberpunk: 'The street finds its own uses for things'[54]. The end result is the replacement of man altogether with a form of artificial life which, though outwardly identical and inwardly capable, is at the same time thoroughly artificial. These then are the replicants of Ridley Scott's *Blade Runner*.

Android dreams – Ridley Scott's Blade Runner

The 1982 *Blade Runner* was supremely influential during the Computer Age. Cited by almost everybody associated with the Cyberpunk movement, and by many scientists, it was directed by Ridley Scott from a screenplay loosely based on the 1968 Philip K. Dick novel *Do Androids Dream of Electric Sheep?*

It is a techno film noir set in a dystopian Los Angeles of 2019. It features a raincoat-clad cop, played by Harrison Ford, pursuing six of the dangerous new Nexus-6 replicants, bio-robotic manufactured beings manufactured by the Tyrell Corporation. They are dangerous because they fuse a lack of any empathy with superior strength and agility. Their Achilles heel is a built-in four year obsolescence. They have travelled to Earth to try to extend their lifespans only to be variously confronted by Ford's Deckard, the eponymous Blade Runner, a specialist cop who hunts down and 'retires' these rogue cyborgs.

The replicants are outwardly indistinguishable from humans, and only a specialized Voight–Kampff test, a polygraph type machine measuring empathic responses, can discover the simulacra. Deckard travels to the Tyrell Corporation's headquarters to see if the test even works on the latest Nexus-6 model.

There he meets Rachael, a femme fatale Nexus-6 with artificial memories implanted, who truly is indistinguishable from human. Only when told of the truth does Deckard realize what she is, too late to stop falling in love with her. *Blade Runner* is famous in science fiction for the fact that it has gone through seven versions of the film, the director's cut famously implying that Deckard himself is a replicant. It also features the much quoted deathbed soliloquy of the replicant leader, Roy Batty, played by Rutger Hauer: 'I've seen things you people wouldn't believe. Attack ships on fire off the shoulder of Orion. C-beams glittering in the dark by the Tannhauser Gate. All those moments will be lost in time, like tears in rain. Time to die'[55].

The stoicism exhibited by Deckard in the face of Batty's emotion lends an interpretation which does indeed, as the director's cut implies, make Deckard the 'sushi'[56] – the cold fish unemotional artificial life form, instead investing Batty with what it means to be human. Batty, the criminal and rebel, feels more in favour than the cop Deckard. In a cybernoir world which has become isolated by the corporate capitalist reductionism of everything to commodity, it is left to those who stand simultaneously both inside and outside that society to identify what it means to be human. By virtue of their artificiality the replicants stand outside the human community, and yet by being one of the commodities they are a component of that society. It is left to them to identify the truth of human existence – that life is transitory: 'the replicants here, especially the lovers Pris and Roy Batty, seem more animated, more beautiful, and more attached to life than their human counterparts'[57]. It is the simulacra who are both more sincere and more serene that those that pursue them. The simulacrum has become more real than the original.

Scientific simulacrum

The notion of a simulacrum can be traced back to Plato, who used it to explain the notion of a false copy of something. It has been argued, most notably by the French Theorist Jean

Baudrillard, that films like *Blade Runner* and books like *Neuromancer* are not set in an alternative to our own world but a simulation, a copy, of the present one. Baudrillard writes in the tradition that starts with Descartes. Dubbed the 'founder of modern philosophy' and the 'father of mathematics', Descartes made the famous assertion *cogito ergo sum* (I think therefore I am), which only allows for consciousness to exist. Everything else could be a simulation, a dream.

Baudrillard argues that the Computer Age made possible the substitution of the simulation for the reality we inhabit. The way we interact with the world around us, by cell phone, Internet and information mediated through the television, means that we often perceive the simulations – film, TV and computer games – as real. We have lost our connection with what is genuine, what is authentic, and have substituted it with mass-produced cloned identities. By wearing the same brand labels we have become ourselves the replicated products of the Computer Age. We have given the Terminator real-world political power, or at least we have given Arnold Schwarzenegger political power, because of our inability to distinguish between the simulacra and the real.

Over 40 years ago, Argentine author Jorge Luis Borges wrote a short story *On Exactitude in Science* about a kingdom so obsessed with mapmaking that their College of Cartographers fashioned a map of the empire as large as the country itself[58]. The useless map was allowed to fray, and its legacy came to be the ruination of cartography. Originally a cutting comment on imperialism and scientific imagination, it is transformed by Baudrillard into an allegory of simulation. A grieving populace bewail the loss of the map, having long occupied it – the simulacrum of the empire – as the real empire. Under the manuscript the authentic area has descended into a desert, a 'desert of the real'[59]. In place of reality, there is a *simulacrum*: 'this map becomes frayed and finally ruined, a few shreds still discernible in the deserts – the metaphysical beauty of this ruined abstraction, bearing witness to an imperial pride and rotting like a carcass, returning to the substance of the soil, rather as an aging double ends up being confused with the real thing'[60].

Baudrillard's notion of hyperreality, that moment when we lose our ability to distinguish reality from fantasy and engage with the fantasy without realizing, can be found clearly in many science fictional texts. The nature of a hyperreal world is characterized by an enhancement of reality; 'Have you ever had a dream, Neo, that you were so sure was real? What if you were unable to wake from that dream? How would you know the difference between the dream world and the real world?'[61].

Walking out of the cinema, moviegoers sometimes experience a sense of dislocation, what was real, what they saw or where they were. This disjunction was explored in Fassbinder's *Welt am Draht* (1973) and the first film version of D. F. Galouye's *Simulacrum 3* (1964), which was remade in 1999 as *The Thirteenth Floor*. Although released in the same year and addressing similar themes it was the Wachowski brothers' *The Matrix* (1999) which was the more popular, and brought the idea of the hyperreal into our homes and hearts.

The Matrix

Baudrillard's influence on *The Matrix* is explicit, as near the beginning of the film the protagonist, Neo, hears a knock at the door in his apartment. At the door is an acquaintance who wants some data, and the data disk is hidden in a book with a hole cut inside it. That book is Baudrillard's *Simulacra and Simulation*.

From that beginning we learn that the hacker Neo, real name Thomas Anderson, has been searching for the answer to the question 'What is the matrix?' Pursued by the forces of 'justice', Anderson is a criminal, a rebel. He is led by an enigmatic woman, Trinity, to meet the equally mysterious Morpheus and his crew, where Neo learns that his life is an illusion played out in a simulation of late twentieth century existence. The reality is that it is actually 2199, and following a man–machine war mankind has become enslaved via this virtual environment, the eponymous

Matrix, and that only the free human city of Zion remains out of machine control. Freed from the control of the Matrix, Neo re-enters the once obligatory artificial environment with his friends. This time, as free agents, they fight with the electronic servants of the machines using skills and techniques downloaded directly into their cerebral cortex.

In the process it is revealed that Neo is 'the one' – the prophesied leader of the humans who will free the humans. The film finishes with Neo successfully mastering his powers and defeating the servants of the machines, only to turn his eyes on the Matrix itself:

> I'm going to show these people what you don't want them to see. I'm going to show them a world without you... a world without rules and controls, without borders or boundaries. A world... where anything is possible. Where we go from there is a choice I leave to you[62].

Neo transcends his reality to become a superhero, a champion, and in the process gains the love of Trinity and a bright future.

The late Computer Age saw a move towards explorations of the idea of transcendence. A popular idea in the science fiction of the time, this was fostered by the burgeoning worship of the technology which spawned it. Transcendence, from the Latin meaning 'to climb beyond', bestows the idea of becoming more than we already are, of exceeding our own limitations. It is found strongly in Gibson's trilogies, in *Blade Runner* and of course in *The Matrix*, and when the Cyberpunk writer Bruce Sterling was questioned as to why so many science fiction novels ended with their protagonists transcending their circumstances, abilities or bodies, he was dismissive:

> It's just a riff, the element of transcendence is just a feature of the SF genre, like feedback in rock music. People who take that stuff seriously end up turning into trolls.... H. P.

Lovecraft was a big fan of that cosmic-type stuff. That may be okay for him, but from the outside what you see is this pasty-faced guy eating canned hash in the dim corner of a restaurant, hands trembly, and a gray film over his eyes'[63].

Despite Sterling's indifference, the idea that 'technology always possesses a pseudoreligious quality. Technology, it is implied, will somehow allow us to transcend our ordinary, human selves'[64] was a strong motif of the Computer Age. This motif was transferred wholesale to the science fiction of the time. The transformative nature of the computer revolution left no doubt as to what we worshipped. As white-robe-clad scientists ordained themselves as the high priests of the Age, an oracle, a source of all information was starting to be installed in every home in the Western hemisphere. The computer became more than just a tool – it became a worldview. Using it for the first time was an almost religious experience. We started to integrate it into every aspect of our lives, to become its disciples. Not just physically implanting it in our bodies as cyborgs, we also surrounded ourselves with the LED and microchips. Consumer electronics became a measure of our status within society. The Reaganomic capitalism of the Computer Age had finally triumphed and we had succumbed to the pressure.

The Matrix itself became a symbol of this. Within a short space of time you could buy the cell phone and you could watch the two sequels and nine animated short films. You could purchase the three computer games and immerse yourself in an online gaming environment all being created to follow the story of Neo and the rest of the human resistance and all generating more income by selling an idea that you could transcend your way out of the dreary by buying the sublime.

Why were The Matrix trilogy and its spin-offs so popular, so successful? The liberation which viewers/gamers/participants felt when they entered the Matrix was intoxicating. Here was an environment where there are no physical limitations to what can be achieved. The characters which inhabit this Universe perform spectacular physical manoeuvres which, although seemingly impossible, are nevertheless achieved with a stylish mixture of panache and self-deprecation. It took the martial arts manoeuvres of the *wuxia pian* Hong Kong hero films[65] and mixed it with the animated stylings of both manga and anime, seasoning it with metaphysical musings. The lesson *The Matrix* attempts to teach is that we are only bound by the limitations of our own mind. Or as *The Matrix* puts it: 'there is no spoon'[66]. Form becomes mutable as those taking part in the Matrix experience take simultaneous journeys: both the physical and the metaphysical fuse. The Matrix experience takes the rhetoric of the Computer Age and hypes it beyond recognition. The notion of liberation from the mundane which the ubiquitous nature of computers promised had not been realized. Like Keanu Reeves' Thomas Anderson at the beginning of the first film, most people who used computers actually felt enslaved to them. Cooped in cubicles, they longed to be able to soar above the clouds as Neo does, to transcend the humdrum realities of their lives. These office revolutionaries desperately want to be at the vanguard of the Computer Age fomenting revolution. They seek to become freedom fighters just like Neo, Trinity and Morpheus. The quixotic nature of the games focuses on the spirit of rebellion. Most computer users never engage in any serious criminal activity, yet within the Matrix one was free to assassinate the forces of law and order in the name of freedom.

Unlike the ultimately capitulatory Winston Smith, caught in the information machines of Orwell's *1984*, the faux rebellious champions of Cyberpunk want to feel capable of threatening the power base of the forces of techno-science. Ensconced in their cosy online communities they prophesied the potential liberating power of the web and a transfer of political power to the people. Yet it was years before the dialectic of blog power and the spectacle of Google censoring itself for China actually started to create any real movement. The battle for control of cyberspace did begin in science fiction.

Francis Bonner's comments about the science fiction of the Computer Age – that it featured a 'concentration on computers, corporations, crime and corporeality – the four C's of Cyberpunk'[67] – seem equally capable of being applied the rest of life at that time. The scientific advances of the computer industry, driven by large corporations like Intel, Microsoft and Apple, drove science fiction to extrapolate a hacker future in which the criminal was the hero. Yet the only way that reality could be transcended and these heroes ever have a chance of winning was by changing the rules of the game completely. Thus the corpo-realities of *The Matrix* and its ilk transformed the structure of the science fictional societies so that these anti-heroes could triumph. Cyclically it was then science fiction which in turn drove the information dissemination process which started with the Internet and ended in the explosion of the World Wide Web. The individualistic nature of science fiction inspired the ordinary corporate drones to become cyberwarriors.

References

1. Advertisement in *Wall Street Journal*, 31 July 1979, pp. 24–5.
2. Broad, W. J.(1992) *Teller's War*. Simon & Schuster, New York, p. 297.
3. Wiener, N. (1948) *Cybernetics, or Control and Communication in the Animal and the Machine*. MIT Press, New York.
4. Cavallaro, D. (2000) *Cyberpunk and Cyberculture*. The Athlone Press, London, p. 14.
5. Davies, M. (1993) Apocalypse soon. *Artform*, December, p. 10.
6. McQuire, S. (2002) Space for rent in the last suburb. In: *Prefiguring Cyberculture: An Intellectual History*. MIT Press, Cambridge, MA, p. 167.
7. Gibson, W. (1984) *Neuromancer*. Grafton, London, p. 9.
8. *Ibid.*, p. 67.
9. *Ibid.*, p. 12.
10. *Ibid.*, pp. 68–9.
11. Interview with William Gibson, CNN, 26 August 1997.
12. Hyman, A. (1982) *Charles Babbage, Pioneer of the Computer*. Oxford University Press, Oxford, pp. 142–3.

13. Sterling, S. (1994) *Mirrorshades – The Cyberpunk Anthology*. HarperCollins, London, p. xii.

14. Hyman, A. (1982) *Charles Babbage, Pioneer of the Computer*, Oxford University Press, Oxford.

15. Singh, S. (2000) *The Code Book: The Secret History of Codes and Code-breaking*. Fourth Estate, London.

16. Dyson, G. (1997) *Darwin Among the Machines*. Allen Lane, The Penguin Press, London, p. 38.

17. Spufford, F. (1996) The difference engine and the difference engine. In: *Cultural Babbage – Technology, Time and Invention* (eds. F. Spufford and J. Uglow). Faber & Faber, London, p. 283.

18. Gibson, W. and Sterling, B. (1990) *The Difference Engine*. Victor Gollancz, London, p. 126.

19. Spufford, F. (1996) The difference engine and the difference engine. In: *Cultural Babbage – Technology, Time and Invention* (eds. F. Spufford and J. Uglow). Faber & Faber, London, p. 288.

20. Sterling, S. (1992) *The Hacker Crackdown: Law and Disorder on the Electronic Frontier*. Bantam Dell, New York, p. 1.

21. Maney, K. (2005) Tech titans wish we wouldn't quote them on this baloney. *USA Today*, 5 July.

22. *Ibid.*

23. *Ibid.*

24. *Ibid.*

25. Kahney, L. (2003) *Grandiose Price for a Modest PC*. http://www.wired.com/, 9 September.

26. Interview with William Gibson, CNN, 26 August 1997.

27. Friedman, E. (1997) *Apple's 1984: The Introduction of the Macintosh in the Cultural History of Personal Computers*. Paper presented at the Society for the History of Technology Convention, Pasadena, California, October, p. 1.

28. Stein, S. R. (1997) *Redefining the Human in the Age of the Computer: Popular Discourses, 1984 to the Present*. University of Iowa Press, http://www.uiowa.edu/~commstud/adclass/1984_mac_ad.html.

29. *Ibid.*

30. Taken from the docudrama *The Pirates of Silicon Valley* (1999), directed by Martyn Burke, TNT.

31. *Ibid.*

32. Friedman, E. (1997) *Apple's 1984: The Introduction of the Macintosh in the Cultural History of Personal Computers*. Paper presented at the Society for the History of Technology Convention, Pasadena, California, October, p. 2.

33. Moore, G. E. (1965) Cramming more components onto integrated circuits. *Electronics*, **38**(8), 2–3.
34. Kurzweil, R. (2003) After the singularity. In: *The Ray Kurzweil Reader*, KurzweilAi.Net.
35. Kurzweil, R. (2001) *The Law of Accelerating Returns*. KurzweilAi.Net, p. 1.
36. Goswani, A. (1983) *The Cosmic Dancers*. McGraw-Hill, London, p. 4.
37. Vinge, V. (1999) in *Vernor Vinge, Online Prophet* by Andrew Leonard, thesalon.com, 5 April.
38. Vinge, V. (1982) *True Names*. Bluejay Books, New York.
39. Heally, E. (2001) in *Waking Life*, dir. Linklater.
40. *Ghost in the Shell* (1995) dir. Mamoru Oshii.
41. Greenia, M. (2003) *The History of Computing*. Lexikon Services, London.
42. Gere, C. (2002) *Digital Culture*, Reaktion Books, London, p. 180.
43. French, S. (1996) *BFI Modern Classics – The Terminator*. British Film Institute, London, p. 7.
44. Bukatman, S. (1993) *Terminal Identity – The Virtual Subject in Post-Modern Science Fiction*. Duke University Press, Durham, NC, p. 301.
45. *Terminator 2: Judgement Day* (1991) dir. Cameron.
46. *Ibid.*
47. Haraway, D. (1991) *Simians, Cyborgs, and Women: the Re-invention of Nature*. Routledge, London, pp. 149–50.
48. Bernal, J. D. (1969) *The World, the Flesh, and the Devil*, 2nd edn. Indiana University Press, Bloomington, IN.
49. Clynes, M. and Kline, N. (1960) Cyborgs in space. *Astronautics*, September, p. 30.
50. Nixon, N. (1992) Cyberpunk: preparing the ground for revolution or keeping the boys satisfied? *Science Fiction Studies*, #57, **19**(2).
51. Cadigan, P. (2002) Introduction to *The Ultimate Cyberpunk*. ibooks, New York.
52. Haraway, D. (2004) A manifesto for cyborgs: science, technology, and socialist-feminism in the 1980s. In: *The Haraway Reader*. Routledge, London.
53. Opening narration from *The Six Million Dollar Man*, 1973–78. ABC Network.
54. Gibson, W. (1986) *Burning Chrome*. Victor Gollancz, London, p. 195.
55. *Blade Runner* (1982) dir. Scott.

56. *Ibid.*
57. Telotte, J. P. (1995) *Replications – A Robotic History of Science Fiction Film*. University of Illinois, Urbana, IL, p. 153.
58. Borges, J. L. (1979) Del rigor en la ciencia. In: *A Universal History of Infamy* (transl. Di Giovanni). Plume.
59. Baudrillard, J. (1988) *Simulacra and Simulations* in *Selected Writings* (ed. Mark Poster). Stanford University Press, Palo Alto, CA, p. 166.
60. *Ibid.*
61. *The Matrix* (1999) dir. Wachowski brothers.
62. *Ibid.*
63. Sterling, B. (2005) quoted in Pontin, J. Against transcendence. *Technology Review*, February.
64. Pontin, J. (2005) Against transcendence. *Technology Review*, February.
65. Williams, W. J. (2003) Yuen Woo-Ping and the art of flying. In: *Exploring the Matrix* (ed. Karen Haber), Simon & Schuster, New York.
66. *The Matrix* (1999) dir. Wachowski brothers.
67. Bonner, F. (1992) Separate development: cyberpunk in film and TV. In: *Fiction 2000: Cyberpunk and the Future of Narrative* (eds. George Slusser and Tom Shippey). University of Georgia Press, Athens, GA.

Chapter 7

THE FRANKENSTEIN CENTURY: THE AGE OF BIOLOGY

In a certain gallery of the Royal Museum in Edinburgh there is an exhibit labelled 6LL3. The product of millions of years of evolution, the exhibit is a fine looking example of the female *Ovis aries* genotype, Finn Dorset variant. A woolly ruminant quadruped that probably descends from the wild mouflon of south-central and south-west Asia, its coat has been used for millennia to provide warmth and comfort for those who domesticated it. Yet this particular example is not just the product of Darwinian evolution; it is also the product of genetics.

Cloned via somatic cell nuclear transfer from a mammary cell in 1996, this is the world's first cloned mammal. Dolly the sheep, named after the prodigiously mammiferous singer, Dolly Parton, stands as a symbol of the Frankenstein century; a hybrid of nature and nurture. Munificently, her eyes address the visitor in kind anticipation, for her appearance never fails to provoke a reaction. An unlikely icon of both science and fiction, Dolly now stands above controversy...

From Darwin to double helix

Earth is an alien planet. It has been for some time now. The paradigm shift of the Copernican revolution cut both ways. Not only did Copernicus make earths of the planets, he also brought the alien to Earth. The Universe of his ancestors had been small, static and Earth-centred. It had the stamp of humanity about it.

Constellations bore the names of earthly myths and legends, and a magnificence that gave evidence of God's glory.

The new Universe was inhuman. The further out the telescopes probed, the darker and more alien it became. 'The history of astronomy', suggests Welsh novelist Martin Amis, 'is a history of increasing humiliation. First the geocentric universe, then the heliocentric universe. Then the eccentric universe – the one we're living in. Every century we get smaller. Kant figured it all out, sitting in his armchair.... The principle of terrestrial mediocrity'[1].

The American astronomer and science fiction writer Carl Sagan had gone further[2]. Sagan saw that humans had suffered a series of 'Great Demotions' in the last five centuries. First there was Earth. It was not at the centre of the Universe. Nor was it the only object of its kind, made of a unique material only to be found on *terra firma*. Next came the Sun. Not at the centre of the Universe, not the only star with planets, nor eternal.

There were more stars in the Universe than grains of sand on all of Earth's beaches. The Milky Way galaxy too proved neither at the centre of the cosmos, nor the only galaxy within it. A hundred billion other galaxies were also discovered; adrift in an expanding Universe so immense that light from its outer limits takes longer than twice the age of the Earth to reach terrestrial telescopes... and there may be other universes. The final demotion, Sagan suggested, would be the discovery of another biological intelligence in the Universe.

If Copernicanism wasn't bad enough, there was Darwin. Man among the microbes, with no special immunity from natural law, and vanishingly little evidence of a divine image. Each successive demotion to date has had an immense impact, both on the human condition and on the meaning of life in the Universe.

With techno-science now unravelling the human genome, the twenty-first century may hold even greater change. We face a future of conscious intervention in reproduction, heredity and directed evolution. Little surprise that this brave new world is dubbed the Frankenstein century.

In its illustrious past, from *Somnium* to cyborg, the themes and icons of fiction became manifest in all areas of society and culture. It's time to consider how science fiction will confront the accelerating pace of change in late capitalist society. This final chapter looks to the past to divine the future. The Age of Biology explores the imaginative way in which fiction has considered biological science, from Darwin to the double helix.

Evolving the post-human

Ever since Copernicus, the post-human has fascinated science fiction, raising questions as to what will become of man, what will become of life in the Universe, and what, if any, is the meaning of life in this new Universe. Since Darwin, thought-provoking science fiction has focused on two crucial developments in modern biology: evolution and genetics.

On the one hand, there have been compelling projections of man's evolutionary future based on Darwin's theory. On the other hand, fiction has been gripped by the remaking of man, the awesome potential of genetics. The finest science fiction has provided a sustained, coherent and often subversive check on the contradictions of science, the promises and pitfalls of progress through the ages. So it has been with biology. Fiction has calculated the human cost of the darker aspects of advances in natural science.

Before falling headlong into the evolutionary future of the post-human, it's worth looking once more at how science fiction operates. Science fiction was described earlier as a response to the cultural shock of discovering man's marginal position in an alien Universe. It works by conveying the taste, the feel, and the human meaning of the discoveries of science. Science fiction is an attempt to put the stamp of humanity back onto the Universe – to make human what is alien.

Religion, like science fiction, is also concerned with the relationship between the human and the nonhuman. Or, more specifically, between the human and the divine. So, as a response to the demotions of life in the Universe, science fiction can be seen, by at least one of the current authors, as a displacement of religion. In this way science fiction may be viewed as the 'soul' of science. Its focus is the human–nonhuman opposition.

With characteristically black humour, Kurt Vonnegut paid tribute to the role of science fiction in his wonderfully creative anti-war novel, *Slaughterhouse Five* (1969). The main character, recently witness to the firebombing of Dresden, considered that people were 'trying to re-invent themselves and their universe. Science fiction was a big help.'[3] Vonnegut himself used science fiction to confront the growing horrors of the twentieth century. Perhaps if his readers could stand the unreality of science fiction, they could face a little bit more reality after reading *Slaughterhouse Five* than they could before they read it: 'everything there was to know about life was in *The Brothers Karamazov*, by Fyodor Dostoevsky. But that isn't enough any more.'[4]

Space, time, machine and monster

Science fiction seems to present an infinity of nightmares and visions, a bewildering diversity of contrasting elements: aliens and time machines, spaceships and cyborgs, utopias and dystopias, androids and alternative histories[5]. But on a more thoughtful level there are four conceptual themes: *space*, *time*, *machine* and *monster*[6]. Each of these themes is a way of exploring the relationship between the human and the nonhuman. Taking a closer look at these themes will enable a clearer understanding of the way in which the genre functions, leading to a more in-depth discussion of biology.

- ◼ *Space*. The science fiction of the space theme represents the nonhuman as some facet of the natural world, such as vast interstellar spaces or the alien, which can be seen as an animated version of nature.
- ◼ *Time*. This theme portrays a flux in the human condition fashioned by some process that is revealed in time. Tales on time

often focus on the dialectic of history, so they are of particular relevance to biology. For example, the evolutionary fable *Last and First Men*, written by Olaf Stapledon, conjures up a post-human *homo superior* in future history.

- *Machine*. These are stories that deal with the man-machine motif, including robots, computers and artificial intelligences (AI). Dystopian tales are part of the man-machine theme; it is the *social machine* in which the human confronts the nonhuman in such cases[7]. Huxley's *Brave New World*, for example, is a hedonistic and ironically ambiguous biotech utopia of repro-technology and social engineering.
- *Monster*. Stories that feature the nonhuman in the form of mutant or monster situated within humanity itself. In these tales there is often an agency of change, such as a nuclear catastrophe, which leads to the change of human into nonhuman. It is within this theme that the remaking of man through genetic design is often encountered. Of course, monsters can be upbeat too, as the countless cases of supermen testify.

This way of thinking about science fiction, as the human versus the nonhuman, is satisfyingly elegant and transparent. Professor Mark Rose deserves credit here[8], since his scheme splendidly serves the purpose of charting science fiction's ongoing dialogue with science (Figure 7.1).

At times, with films like *Deep Impact* (1998) and *Armageddon* (1998), science and the human are pitched against nature and the nonhuman. In these cases, the nonhuman comes in the form

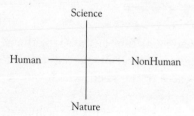

Figure 7.1 The paradigm of science fiction, after Professor Mark Rose[9].

of a rogue comet or asteroid, shattering the stability of the human world. In dystopias, such as *Brazil* (1985) and *The Matrix* (1999), nature and human are united in opposition to science and non-human. As these dystopias suggest, science fiction may characterize science as nonhuman and unnatural. In *The Matrix*, for instance, the natural and organic human homeworld of Zion counters the mechanical and scientific alternative world of the Matrix itself. According to this convention, utopias are imagined societies that are more fully human than the present.

More often, though, science features on both sides of the human–nonhuman conflict. In *The War of the Worlds*, for example, science is part of the nonhuman element symbolized by the invading Martians. They are agents of the void. They also embody science with their vast, cool and unsympathetic intellects. Later, however, these alien invaders fall victim to microbes, a fate which pits the science and understanding of unsolicited natural selection on the side of the invaded humans.

Anticipations of the coming race in space and time

With this thematic *space*, *time*, *machine* and *monster* view of science fiction, a history can be told of the hopes and fears for biology; a story that spans all ages, from the Age of Discovery to the Frankenstein century. It is a story of the ongoing relationship between science and the cultural scepticism of its fiction. Indeed, science fiction, as a form of the fantastic, has been the *literature of change* since the Renaissance. It has provided a commentary on the accelerating pace of change in capitalist society, and continues to use the fantastic to make sense of the dark magic of nature.

The publication in 1859 of Darwin's *Origin of Species* had a tremendous impact on the popular imagination. It radically challenged traditional ideas of human nature, purpose, and relationship to God. Its majestic reach embraced questions of morality, economics and political policy. Social Darwinism arrived. Nationalists used Darwinism to argue for a strong state as the fittest among nations, militarists found in it the sanction for war, and imperialists identified the motive for the conquest of 'inferior races'.

In his forceful history of science fiction, *Billion Year Spree* (1973), Brian Aldiss bears witness to the surfacing of the 'submerged nation' theme in Victorian fiction. *Sybil* (1845), written by British Prime Minister Benjamin Disraeli, had first manifested the idea that British society comprised two distinct nations, the Haves and the Have-nots. 'A grave moral division lay at the basis of Victorian hypocrisy', suggested Aldiss[10]. Guilty consciences had inspired science fiction writers, such as Bulwer-Lytton and H. G. Wells, to use Darwinism to expose such fears. The oppressed (the fictional 'submerged nation' races of the Morlocks and the Vril-ya) would have their day in the Sun.

Some rejected Darwinism out of hand. Disraeli drew his battle line[11]:

What is the question now placed before society with glib assurance the most astounding? The question is this – is man an ape or an angel? My Lord, I am on the side of the angels[12].

Dream on, Disraeli. By the 1880s the battle was over. Darwinism was the intellectual creed of the age, and has lit up the conceptual sky ever since. Darwin's theory gave credence to the development of life, and not just under Earthly conditions. It also raised the possibility of physical cosmic evolution. The rise of spectroscopy transformed astronomy into astrophysics[13]. Here was evidence that natural law governed right through the Universe. Like Copernicanism, Darwin's theory transfused lifeblood into extraterrestrial life. It revolutionized the cosmic perspective, suggesting that life was a basic property of the Universe.

So it's hardly surprising that science fiction uses the *space* and *time* themes to first explore the question of the post-human. French astronomer Camille Flammarion was evangelical in his support for extraterrestrial life, exercising a great influence on twentieth century attitudes to the idea[14,15]. Just three years after the publication of Darwin's theory, Flammarion released his *La Pluralité des Mondes Habités* (Plurality of Inhabited Worlds, 1862). Over the next twenty years, 33 editions of *La Pluralité* were published, a clear indication of the popularity of both Darwinism and the idea of extraterrestrial life. Flammarion argued, with enthusiasm, that alien life, originating spontaneously rather than divinely, evolved through natural selection in its extraterrestrial setting.

Anthropocentrism was cast out. Planet Earth and its inhabitants were relegated to a lowly rung on the evolutionary ladder. It was an idea that would come to dominate twentieth century film and fiction, explored through the works of H. G. Wells, Olaf Stapledon and Arthur C. Clarke.

Close encounter of the first kind: H. G. Wells

In the late nineteenth century biology finally found its way into fiction. Notwithstanding the development of the idea of the post-human since Copernicus, no one could have predicted that evolution would spark one of the universal motifs of twentieth century fiction: the concept of the alien.

As a result, an increasing number of people met the ideas of Darwin, not through science, but as a text, inspiring emotional as well as intellectual reactions. In this way, the concepts of evolution and the future of man were embedded ever deeper into the public psyche. It's worth remembering that the creative morphing of scientific ideas into symbols of the human condition:

is often an unconscious and therefore particularly valuable reflection of the assumptions and attitudes held by society. By virtue of its ability to project and dramatise, science fiction has been a particularly effective, and perhaps for many readers the only, means for generating concern and thought about the social, philosophical and moral consequences of scientific progress[16].

Scientists are creatures of the culture in which they swim. Alien contact narratives motivated a significant number of such scientists. The idea of life in the Universe, and man's place within it, was firmly fixed in the scientific as well as the popular imagination. Darwin's theory had given credence to the evolution of life on Earth and to evolution in a cosmic setting. Darwin inspired a wealth of fiction[17] and provided a rationale for imagining what cosmic life might develop. From now on the idea of cosmic life became synonymous with the physical and mental characteristics of the alien. It provided a rubric against which man himself could be measured.

The impact of Wells' fiction was colossal. *The Time Machine* (1895) and *The War of The Worlds* (1898) were responsible for igniting both the *space* and *time* themes in the genre of science fiction and in the public imagination. Wells created the nexus of the alien, armed with its potential for probing human evolution. Wells' early books 'are, in their degree, myths; and Mr Wells is a myth-maker'[18].

Once developed by Wells, the alien idea proved a potent motif for cultivating fictional explorations of the singularity or insignificance of humanity. During such explorations, the secondary question of the character of alien and interspecies interaction became an issue, which later affected the Search for Extraterrestrial Intelligence (SETI) science programme. As Brian Aldiss put it, 'Wells is the Prospero of all the brave new worlds of the mind, and the Shakespeare of science fiction'[19].

Close encounter of the second kind: Olaf Stapledon

If H. G. Wells is the Shakespeare of science fiction, then Olaf Stapledon is its Milton. A philosopher based at the University of Liverpool, Stapledon used the genre to explore nothing less than the meaning of human existence in a cosmic setting. His two key works, *Last and First Men* (1930) and *Star Maker* (1937), opened up emergent philosophical and spiritual issues through science fiction.

In the preface to *Last and First Men* Stapledon tells the reader that his story is an attempt 'to see the human race in its cosmic setting, and to mould our hearts to entertain new values'[20]. In a telling evocation of Darwin's theory, he suggests that such attempts to extrapolate man's evolutionary future 'must take into account whatever contemporary science has to say about man's own nature and his physical environment'[21]. Stapledon produced a fiction that incorporated the most recent ideas of astronomy and evolutionary biology. He synthesized a new form of myth apposite to a scientifically cultured twentieth century. In the words of Stapledon himself, the aim must not be just 'to create aesthetically admirable fiction... but myth'[22].

Francis Bacon had laid the basis for a militant, aggressive science – a science that actively penetrated the natural world for the relief of man's estate. By the early decades of the twentieth century, an even more radical scenario emerged: the improvement of man's estate might best be realized by the biological upgrading of man himself. In 1927 it was discovered that the mutation of reproductive cells could be greatly increased through exposure to X-rays.

Last and First Men

In the emerging world of genetics, Stapledon imagined the future forms of man. *Last and First Men* is a future history on a staggering scale. The 'hero' of the book is not a man, but mankind. The story embraces seventeen evolutionary mutations, from the present 'fitfully-conscious'[23] First Men, to the glorious godlike Eighteenth Men, who reign on Neptune. It is a history that spans two thousand million years.

The first and most infamous exponent of the genetic intervention in the human race was Darwin's cousin, Francis Galton. It was he who had introduced the word *eugenics*. In *Last and First Men*, Stapledon thought long and hard about eugenic practices. One of the causes of the demise of the First Men, for instance, was their failure to realize a eugenics program, 'In primitive times the intelligence and sanity of the race had been preserved by the inability of its unwholesome members to survive. When humanitarianism came into vogue, and the unsound were tended at public expense, this natural selection ceased. And since these unfortunates were incapable alike of prudence and of social responsibility, they procreated without restraint, and threatened to infect the whole species with their rottenness'[24]. So human intelligence steadily declined, 'And no one regretted it'[25].

Later came the irresistible rise of the Third Men. With their rediscovery of eugenics, the Third Men focused their efforts on that most distinctive feature of man, the mind. Seeking to 'breed strictly for brain, for intelligent coordination of behaviour'[26], the climax of their engineering program was the Great Brains.

The Great Brains first helped, then enslaved, and finally eliminated their creators. Ultimately they turned their cool intellects upon themselves. They created a superior species, the Fifth Men. The Fifth Men were accomplished in art, science and philosophy, perfectly proportioned of body and mind. They were able to travel mentally back through time to experience the whole of human existence. Indeed, the Fifth Men became the most perfect species ever to dwell on Earth.

Star Maker

The conceptual reach of *Star Maker* beggars belief even further. Its scope is so vast that *Last and First Men* would warrant a mere page in the cosmic sweep of Stapledon's next book. The presence of the alien in *Star Maker* is, again in the words of Stapledon himself, to 'explore the depths of the physical universe [and] discover what part life and mind were actually playing among the stars'[27].

Stapledon was twenty years ahead of the game. The setting in which *Star Maker* was conceived had undergone a further, though more silent, revolution in cosmology. It was not until the late 1950s that astronomers drew analogies between revolutions in cosmology and the impact of finding extraterrestrial intelligence[28]. It was suggested[29] that alien contact would represent

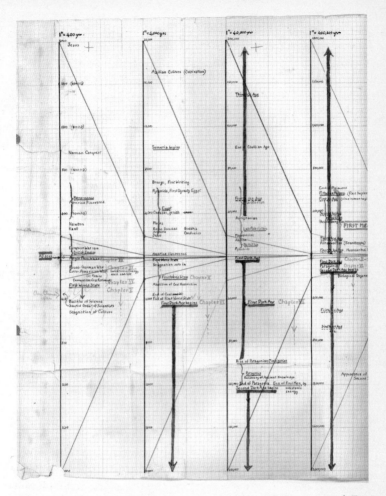

Figure 7.2 Olaf Stapledon's timechart storyboard for *Last and First Men*.

'The Fourth Adjustment' in humanity's outlook, following the shift to the geocentric, heliocentric and 'galactocentric' worldviews. This latter revolution, hastened by discoveries showing that our local solar system was merely at the edge of our galaxy

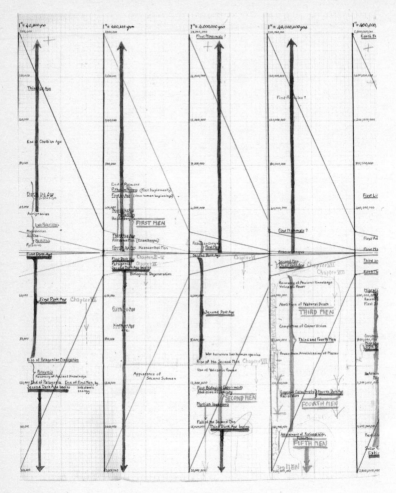

Figure 7.2 (*continued*)

and that the galaxy itself was but one of many, was made just prior
to the time Stapledon was writing *Star Maker*.

Astronomy had undergone great revolutions[30]: the Copernican, the galactocentric, and Hubble's discovery of an expanding

Universe. But there was one massive upheaval yet to come: the answer to the question 'Are we alone in the Universe?'. The revolution had already begun with Stapledon. By the mid-1920s, revolutions, including those of Copernicus, Darwin and Einstein, may have inured the masses to marginalization[31]. Stapledon was preparing the public for the final great demotion, and in the process helped develop the myth of the close encounter of the third kind: physical contact.

In *Star Maker*, alien biologies, together with terrestrials, search for the supreme intelligence in the new Universe. Stapledon's story is a quest for the spirit of the cosmos, an entity at the head of a new, and cosmic, great chain of being. In an early evocation of the implicit inhumanity of the new Universe, he writes 'it was becoming clear to us that if the cosmos had any lord at all, he was not that spirit [God], but some other, whose purpose in creating the endless fountain of worlds was not fatherly toward the beings that he made, but alien, inhuman, dark'[32].

Stapledon's fiction, then, emphasized the triviality of humanity in the face of a new and vast cosmos, which itself may harbour truths and meaning as yet unknown to an immature terrestrial race. His fiction on the question of intellectual contact with alien biologies had great influence on working scientists, such as exobiologist J. B. S. Haldane, physicist Fred Hoyle, Carl Sagan (one of the founders of SETI in the early 1960s), and fiction writer Arthur C. Clarke.

Close encounter of the third kind: Arthur C. Clarke

So, more than science itself, the fictional elaborations of Darwin's theory, through the *space* and *time* themes, defined the pervasive role of the post-human in popular culture. Arthur C. Clarke's 1953 novel *Childhood's End* is an archetype of the way in which alien fiction developed the post-human. Clarke had already written a number of short stories on the alien motif. The influential *The City and the Stars* portrays humanity confronted with extraterrestrial cultures and intelligences 'he could understand but not match, and here and there he encountered minds which would soon have passed altogether beyond his comprehension'[33].

In *Childhood's End* Clarke developed the myth of contact through an alien invasion. The Overlords are benevolently responsible for guiding humanity to an even greater intelligence, the Overmind. Clarke uses the extraterrestrial context to highlight humanity's immaturity in an aged and cultured Universe. The Overlords exact an end to poverty, ignorance, war and self-government. There is a payoff, however. It is a preparation for the final destiny of humanity: Earth's children are sacrificed and united within the collective of the Overmind.

In the words of Clarke himself 'the idea that we are the only intelligent creatures in a cosmos of a hundred billion galaxies is so preposterous that there are very few astronomers today who would take it seriously. It is safest to assume, therefore, that They are out there and to consider the manner in which this fact may impinge upon human society'[34]. *Childhood's End* was written amidst growing claims for inexhaustible exoplanetary systems. It was not until 1995 that empirical evidence for such extrasolar planets was discovered.

The novel, and indeed much of Clarke's fiction, reflects his scientific belief in extraterrestrial life and eventual contact. Interestingly, in the preface of a 1990 reprint and partial rewrite of *Childhood's End*, Clarke attempts to unravel pseudoscience from the extraterrestrial message underlining the original narrative:

> I would be greatly distressed if this book contributed still further to the seduction of the gullible, now cynically exploited by all the media. Bookstores, news-stands and airwaves are all polluted with mind-rotting bilge about UFOs, psychic powers, astrology, pyramid energies[35]...

Clarke is adamant, however, as to the continued relevance of his book:

I have little doubt that the Universe is teeming with life. SETI is now a fully accepted department of astronomy. The fact that it is still a science without a subject should be neither surprising nor disappointing. It is only within half a human lifetime that we have possessed the technology to listen to the stars[36].

Science fiction and SETI

Science started searching for the alien. The emotional question of man's place in the cosmos was woven into all scientific discussions on life in the Universe. Increasingly, the dialectic between anthropocentrism and pluralism was revolutionized time and again by stunning discoveries in physics. Firstly, there was the progress of relativity and quantum theory. Next came the Big Bang model of an expanding Universe.

The idea of a cosmos centred on our unique and privileged position in space–time became more and more preposterous. Instead, science developed the 'cosmological principle'. Based on applying General Relativity to the large-scale structure of the Universe, the principle, simply stated, suggests that there is no special place in the Universe. It is the belief that Martin Amis referred to as the 'assumption of mediocrity'.

In the wake of the massive popularity of contact fiction, SETI pioneer Frank Drake became the first radio astronomer to contemplate the transmission of an alien signal. His Project Ozma in 1960 examined two sun-like stars, Tau Ceti and Epsilon Eridani. The project mission was to locate intelligent signals coming from hypothetical planets orbiting the stars. Project Ozma was followed in 1964 by project CETI (Communication with Extra-Terrestrial Intelligences), which greatly influenced the design of listening programmes using the largest radio telescopes on Earth.

In a seminal conference at Green Bank, West Virginia, in 1961, Frank Drake presented his now famous 'Drake equation'. The equation quantifies the factors that may determine the number of communicating civilizations in our Galaxy. The publication in

1966 of *Intelligent Life in the Universe* by Carl Sagan and Russian astrophysicist Iosif Shklovskii was another telling advance. It was the first seminal scientific text on the question of extraterrestrial intelligence. Sagan later developed a fictional exploration of alien contact in his 1985 novel *Contact* and its cinematic counterpart, released in 1997. Both were grounded in Sagan's experience of the scientific search. They portray humanity's destiny, again among culturally and intellectually superior extraterrestrials.

2001: A Space Odyssey

It was Arthur C. Clarke, however, who was instrumental in developing the alien for the mass market. The imaginative flood and sweep of the alien motif in fiction is impressive enough. As far as the propagation of the idea of the post-human is concerned, however, it was the sway of cinema that beamed the broadcast far further. Cosmic fiction and its exploration of humanity could be examined without the alien. But the inclusion of the alien was evolution's defining moment in the public imagination. In addition, science fiction would have stayed marginalized were it not for the opening up of the genre to film and television.

Clarke and Stanley Kubrick's *2001: A Space Odyssey* (1968) was delivered during the peak of the extraterrestrial hypothesis (ETH), 1966 to 1969[37]. The ETH held that UFOs were close encounters with visiting aliens, a hypothesis vastly influenced by the mythic fiction of Wells, Stapledon and Clarke himself.

Famed for the maturity of its portrayal of mysterious, existential and elusive aliens, *2001* raised science fiction cinema to a new level. The eminent US film critic Roger Ebert, when asked which films would remain familiar to audiences 200 years from now, selected *2001*. Another critic claimed the picture was an 'epochal achievement of cinema' and 'a technical masterpiece'[38]. The film, not the book, made Clarke the most popular science fiction writer in the world. Kubrick's masterpiece, which made dramatic and sophisticated use of the alien premise, quickly became a classic discussed by many, if not understood by all.

Kubrick's bible was *Intelligent Life in the Universe*. Kubrick had originally filmed interviews with 21 leading scientists about the possibility of alien life as a prologue to the film's narrative. Interviewees included physicists Frank Drake and Freeman Dyson, anthropologist Margaret Mead, roboticist Marvin Minsky, and Alexander Oparin, the great Soviet authority on the origin of life, often described as the 'Darwin of the twentieth century'. Kubrick's intention was to lend astrobiology that special dignity it has only acquired since. Though the interviews were cut from the final version of the film, a book of the transcripts was published in 2005.

The ultimate trip

2001 is an epic journey – the 'ultimate trip', as it was billed in those *New Age* counterculture days. Darwin had inspired the German philosopher Friedrich Nietzsche to write *Also Sprach Zarathustra*. The book identifies three stages in the evolution of man: ape, modern man, and ultimately, superman. As Nietzsche put it, 'What is the ape to man? A laughingstock, or painful embarrassment. And man shall be to the superman: a laughingstock or a painful embarrassment'[39]. Modern man is merely a link between ape and superman. For the superman to evolve, man's will, 'a will to procreate, or a drive to an end, to something higher and farther'[40], must power the change.

Likewise, Kubrick's movie traces man's journey though three stages. As the film's subtitle suggests, the narrative is a spatial odyssey from the subhuman ape to the post-human starchild. The unfolding four-million-year filmic story embraces each theme of science fiction: space (contact through alien cultural artefacts), time (evolutionary fable), machine (the man–machine encounter with HAL, computer turned murderer), and monster (human metamorphosis)[41]. The opening 'Dawn of Man' scene of *2001* sees the Sun rise above the primeval plains of Earth, to the rising soundtrack of Richard Strauss's Nietzsche-inspired tone poem, *Also Sprach Zarathustra*. A small band of man-apes are on the long, pathetic road to racial extinction.

The journey begins with one of the hominids exultantly hurling an animal bone into the air. In an astounding cinematic ellipsis the bone instantly morphs into an orbiting satellite, and three million years of hominid evolution is written off in one frame of film. The agency that drives the guided evolution of these early hominids is an alien artefact in the shape of a black monolith. Like Wells' Martians, the monolith embodies the void[42]. Primal bone technology marks the birth of the modern era. Man and machine, from the very outset, are inseparable. The mysterious presence of the monolith transforms the hominid horizon. The journey to superman begins.

The intelligent use of film technology is one of the factors that make *2001* such a *tour de force*. Winner of an Oscar for special effects, the film seemed to offer a more 'realistic' picture of space travel than the endeavours of Armstrong and Aldrin only a year later. Sections of the film were used in training NASA astronauts. Indeed, Arthur C. Clarke later suggested that of all the responses to the film, the one he valued most highly was that of the Soviet cosmonaut Alexei Leonov: 'Now I feel I've been into space twice!'[43]

Man's growing maturity through an early space age now unfolds in a three-way narrative of machine, human and post-human. A space voyage leads to further waystations of the monolith. A black obelisk is uncovered at a pioneering Moon base, and a mission is sent to Jupiter, to where the mysterious lunar artefact seems to be sending its alien signal. The banality and vacuity of the human crew members is sharply and ironically contrasted by the robust intelligence of the ship's onboard computer, HAL 9000. The film was one of the very first to carry 'product placements' for companies such as IBM, Pan Am, and AT&T. Indeed, space travel is replete with corporate logos and trademarks, showing a world

absolutely managed – the force controlling it discreetly advertised by the US flag with which the scientist [Doctor

Floyd] often shares the frame throughout his 'excellent speech'... and also by the corporate logos – Hilton, Howard Johnson, Bell – that appear throughout the space station. In 1968, the prospect of such total management seemed sinister – a patent circumvention of democracy[44].

The irony and satire of the film's portrayal of a bland future dominated by corporations and technology was lost on some. British scientist and software developer Stephen Wolfram said the film's futuristic technology greatly impressed him as a boy. Microsoft co-founder Bill Gates has suggested that 2001 inspired his vision of the potential of computers[45] (though whether Gates was also inspired by the picture of sinister corporate domination is pure speculation). Nonetheless, such corporate control is symptomatic of the spiritual crisis of the early space age portrayed in the film.

The potent evolutionary force imparted by the black obelisks is overdue. The space age was ultimately inspired out of the apes by the alien intelligence. Now, the odyssey of self-discovery culminates under the watchful presence of the monoliths when modern man, in the form of the individual astronaut David Bowman, comes to an end. With the massive presence of planet Earth filling the screen, the foetus of the superhuman star child floats into view. Moving through space without artifice, the image suggests a new power. Man has transcended all earthly limitations.

A 'scientific definition of God'

Stanley Kubrick claimed the film provided a 'scientific definition of God'[46]. There was little drama in Darwin's evolution – just the slow, solid state of inexorable change. So Kubrick and Clarke invoked a fictional form of Stephen J. Gould's 'punctuated equilibrium'. The film augments the usual driving force of evolution, long periods of steady change, with the episodic guiding hand of superior beings. It is a story of the effective creation and resurrection of Man.

As Clarke suggests in his book of the screenplay:

> Almost certainly there is enough land in the sky to give every member of the human species, back to the first apeman, his own private, world-sized heaven, or hell. How many of those potential heavens and hells are now inhabited and, by what manner of creatures, we have no way of guessing; the very nearest is a million times further away than Mars or Venus, those still remote goals of the next generation. But the barriers of distance are crumbling; one day we shall meet our equals, or our masters, among the stars[47].

A final key influence can also be identified. Physical scientists have historically held a deterministic view of the possibility of extraterrestrial life. As the Clarke quote above from *2001* clearly shows, this determinism is based mostly on the physical forces in the Universe – the idea that the sheer number of stars and orbiting planets is statistically sufficient to suggest other Earths lie waiting in the vastness of deep space. Fiction, for many centuries, followed suit. Since Copernicus came before Darwin, and physics before biology, fictional accounts of alien life have usually been positioned firmly in the pro-SETI, pro-life camp of the extraterrestrial life debate. By the twentieth century, an entire generation of future SETI-hunters were cast under the same spell.

As the millennium drew to a close, the story changed. Pioneers of the evolutionary synthesis (the fusion of evolutionary biology with genetics) were Theo Dobzhansky and Ernst Mayr. They emphasized that whilst physics and fiction still think along deterministic lines, evolutionists are impressed by the incredible improbability of intelligent life ever to have evolved, even on Earth. We may, after all, be alone in the Universe. Such has been the power of science fiction. Its exploration of evolution and the future of man led directly to a huge investment in the serious search for ET.

And DNA

Genes are common to all life. They are the method by which any living organism receives common traits from its ancestors. These traits are inscribed on the living ladder of deoxyribonucleic acid (DNA). The blueprints for both the construction and operation of an organism, DNA can be found in every living thing on Earth. How DNA occupies a specific spot in its macro unit of a chromosome determines a specific characteristic of the living thing, and is referred to as a gene.

Charles Darwin's work, which resulted in his theory of natural selection, provided the basis for evolutionary theory. Yet it was only when an Augustinian Abbot, Gregor Mendel, pioneered discoveries in heredity, variation and reproduction that the way in which evolution works in practice was comprehended. Mendel's experiments with peas in the mid-nineteenth century led him to appreciate how heredity in sexual reproduction works. It functions through the integration of separate factors, not just the blending of inherited characters. There is doubt as to the probity of this Jesuit scholar, some claiming that his data was falsified whilst others argue that it is accurate[48]. None can doubt the influence of his research on the world in which we live and in particular the century that we are just beginning to move through, the Frankenstein century.

Not surprisingly, considering it was carried out in a monastery garden, Mendel's work was ignored. Despite his assertion that 'My scientific studies have afforded me great gratification; and I am convinced that it will not be long before the whole world acknowledges the results of my work'[49], it took over forty years before three botanists, all working independently, rediscovered Mendel's contribution and in the process heralded him as the 'father of genetics'.

It was in the Astounding Age, at the end of the First World War, that the British biologist Ronald Fisher[50], hailed by some as 'Darwin's greatest twentieth-century successor as well as the father of modern statistics'[51], started the 'modern synthesis'.

Fisher integrated Charles Darwin's theory of the evolution of species by natural selection with Mendelian genetics. This synthesis was to gain pace throughout the twentieth century. One of its key theoretical contributors was Julian Huxley, grandson of Darwin's bulldog, T. H. Huxley. As well as being a noted evolutionary biologist, Huxley was also a science communicator. Working with H. G. Wells they produced the popular nine-volume piece of science communication *The Science of Life*. Mixing media, he also won an Oscar for his work on the world's first nature documentary, *The Private Life of the Gannets* (1934). His zoological work led to a concern for education and conservation which was recognized when in 1946 he was appointed the first Director-General of UNESCO (United Nations Educational, Scientific and Cultural Organization) and in 1961 to found the World Wildlife Fund (WWF). Yet in the popular history of genetics he is overshadowed by his brother, Aldous.

Not just famous for authoring the original screenplay to Disney's *Alice in Wonderland*, Aldous Huxley is a science fiction icon whose vision of a future defined by genetics was encapsulated in the classic science fiction novel *Brave New World*.

How beauteous mankind is! O brave new world, that has such people in't![52]

Inspired by his brother's colleague and luminary of science fiction H. G. Wells, *Brave New World* (1932) is a negative response to Wells' utopian novel *Men Like Gods* (1923). It defied the optimism of Well's work and embracing a dystopian vision, a vision which Wells himself explored in *The Sleeper Awakes* (1910). *Brave New World* presents a disconcerting picture of twenty-sixth century London.

Aldous Huxley's vision for the future takes account of his own genetic history. The accumulated scientific work of his grandfather was seemingly laid out in the lives of both his brother and he. Huxley literally incarnates the premise that science drives fiction and fiction in turn drives science.

Huxley extrapolates a future in which there is no war, no poverty and no pain, all through the application of genetic science. It may seem like a fitting tribute to his grandfather, yet it harbours dark secrets, for all these strikingly beneficial attributes have been attained by eliminating any genetic variance in the population, and in the process ridding them of all that makes them individuals. A homogeneously hedonistic society, which papers over its cracks by embracing promiscuity, extensive drug use and hallucinatory fantasies, its motto, 'Community, Identity, Stability'[53], hints at its caste-based system – a system which indoctrinates its children:

> Oh no, I don't want to play with Delta children. And Epsilons are still worse. They're too stupid to be able to read or write. Besides, they wear black, which is such a beastly colour. I'm so glad I'm a Beta. Alpha children wear grey. They work much harder than we do, because they're so frightfully clever. I'm really awfully glad I'm a Beta, because I don't work so hard. And then we are much better than the Gammas and Deltas. Gammas are stupid[54].

Divided into three main sections, the book first introduces us to the features of the World State, a society which Huxley based on an early trip to the United States. Huxley's social order marks its consumer-driven, inward-facing, relentlessly cheerful, gum-chewing, feelie-watching nightmare in years since the Ford Motor car was invented. It is introduced to us through the eyes of Bernard Marx, an Alpha, the highest of the social orders. Bernard feels somewhat out of place due to the contamination of his embryo with alcohol during the Bokanovsky process of cloning. Stuttering uneasily through society he gains permission to take his current partner, the 'pneumatic'[55] Lenina Crown, outside the boundaries of the World State to a 'reservation' in New Mexico where savages live.

The second section chronicles the interaction between the civilized visitors and one native in particular, named John the

Savage, the son of a civilized woman. This woman, Linda, was abandoned on the reservation. The Savage's philosophy and notions of social interaction in the civilized world outside the boundaries of the concentration camp are defined by a copy of the *Complete Works of William Shakespeare*, a book from which his mother taught him to read. Perceived by Bernard as an opportunity to gain both status and credibility with other Alphas, he and Linda are transported back to civilization.

The final portion of the book revolves around the culture shock which John's arrival in the world state creates. Alternately happy because he has fallen in love with Lenina and desolate because of the empty sham of a life he perceives around him, John oscillates between embracing the hedonism of the World State and castigating it. While his mother floats away in a drug-induced haze, John debates with an Alpha colleague of Bernard's, Helmholtz Watson. Arguments over Shakespeare's writing provide some respite. After Linda dies, John runs amok and is brought in front of the Resident World Controller for Western Europe, Mustapha Mond. There Mond informs him that he will never be free, his status as an experiment guaranteeing his captivity. Mond meanwhile informs the Alphas, Bernard and Helmholtz, that they are to be exiled to an enclave of the World State reserved for such free spirits as they. The climax of the novel comes when, torn apart by competing desires, John the Savage, like Shakespeare's Othello, commits suicide, his mute swinging body a savage indictment of the onlookers who have gathered to watch the spectacle.

No man is an island[56] – Aldous Huxley

Later Huxley was to explore the issues raised by *Brave New World* from the other side of the coin. His avowedly utopian novel *Island* (1962) is the product of his own changing vision of the *Brave New World* he created. He remarks in the revised introduction to an edition of the novel released in the Atomic Age:

If I were now to rewrite the book, I would offer the Savage a third alternative. Between the Utopian and primitive horns of his dilemma would lie the possibility of sanity.... In this community economics would be decentralist and Henry-Georgian, politics Kropotkinesque co-operative. Science and technology would be used as though, like the Sabbath, they had been made for man, not (as at present and still more so in the Brave New World) as though man were to be adapted and enslaved to them[57].

Thus in his *Island* Huxley rejects the pharmacologically addicted, hedonistic elimination of the individual to instead use narcotics for beneficial uses and to make sex an expression of the individual, not the group[58]. Whilst artificial conception is practised this is done not via rows of bottles, as one after another clones are dispensed from a production line. Instead, in *Island* third party donation is utilized as a means to aid conception. The inhabitants of Huxley's idyll embrace modern science and technology to improve medicine and nutrition, but reject the unabashed pursuit of capitalism so beloved by the World State. Perhaps as much as his *Brave New World*, Huxley's *Island* deserves to be read, for 'the novel's warnings about religious fanaticism, the exercise of massive military power, the geopolitical importance of oil and the development of artificial insemination seem extraordinarily prophetic'[59].

Scientists that callously produced identical consumers in *Brave New World* become elevated to the status of liberators in *Island*. Huxley's work reminds us that the culture of science in the popular imagination has long been defined by the way that scientists themselves have been perceived and genetics is no different. For every set of heroes labouring away in the back rooms of obscure laboratories to emerge blinking into the light there is a foil. Champions like James Watson, Frances Crick and Rosalind Franklin, whose isolation of the double helix structure of DNA led to worldwide recognition and popular acclaim, even being made into a film[60], are rare.

It is the image of the mad scientist that predominates[61]. In genetics the archetypical mad scientist is based on the real-life atrocities of the Nazi 'Angel of Death' Josef Mengele. His exploration of heredity in the Auschwitz and Birkenau concentration camps by experimenting on twins and those with any abnormality led to his elevation to a place in the genetic dark pantheon, a place which rivals those of Victor Frankenstein or Wells' Moreau. He was immortalized in the science fiction film *The Boys from Brazil* (1978) where, played with aplomb by Gregory Peck, he tries to clone Hitler and bring about a fourth Reich, only to be foiled by Laurence Olivier as an aging Nazi hunter. He also appears in *Marathon Man* (1976), this time superficially disguised as Dr Christian Szell, who in a reversal of fortunes is played by Laurence Olivier; he attempts to smuggle diamonds from Latin America to the West whilst engaging in some excruciatingly creative dentistry on Dustin Hoffman.

Mad, bad and dangerous to know[62]

The Mengelian model of a scientist without any concern for consequences is one which the Frankenstein century has embraced. The selfish pursuits of the isolated genius are particularly frightening when applied to the human genome. We may have been afraid of destruction at the hands of Strangelove and his ilk, yet more frightening is the idea that we will be transformed, that we will be twisted by the machinations of science. Films like *28 Days Later* (2002), *Resident Evil* (2002) and *Doom* (2005) feature a post-apocalyptic landscape, the result of genetic mutations released from multinational commercial laboratories. In these movies criminally negligent and homicidally inclined scientists have made their mark in popular culture.

The latter two movies are just the tip of their franchises. Both are based on bestselling first-person shooter video games that have multiple incarnations and numerous spin-offs, and have made millions. In these games and films the unreasoning enemy are your neighbours, friends and family transformed by science into monsters. Mark Rose's conception of the monster intersects here with the scientific machine that has created them. The commercialization of science and the genetic industry in particular have found a voice in the science fiction of the Frankenstein century. The computer game industry has placed the viewer within the nightmare of science gone mad. Armed with a heavy arsenal we are encouraged to deal severely with the mutant creations before confronting the ultimate enemy, the scientists themselves.

Culpability is directed at the scientific establishment in these products. Bad science becomes no longer the provenance of the maverick alone. Carl Sagan's assertion that 'Even Johannes Kepler, Isaac Newton, Charles Darwin, Gregor Mendel, and Albert Einstein made serious mistakes. But the scientific enterprise arranges things so that teamwork prevails: What one of us, even the most brilliant among us, misses, another of us, even someone much less celebrated and capable, may detect and rectify'[63] is dismissed in a hail of automatic gunfire. These films assert the notion that science as a cooperative enterprise is subverted by the military-industrial complex to become both more frightening and more effective.

Nowhere in the science fiction of the Frankenstein century is the collectivization of scientific tyranny more apparent than in the *Bas-Lag Chronicles* of the multiple Hugo award nominee China Miéville. Part of a conscious movement at the beginning of the twenty-first century to transcend the boundaries of literature in science fiction and fantasy, Miéville's work is part of a subgenre known as the *New Weird*. It is controversial because of the hybrid nature of both its style and content; there are also arguments, like those which encompassed the Cyberpunk of the Computer Age, that it is not a movement or an effective grouping. Still, Miéville has identified himself as part of the group[64] and his writing certainly mixes the surreal fantastic with the scientific and extrapolative.

New Weird – China Miéville

Miéville's trilogy is set on a world which features numerous different races sitting alongside each other, its principal protagonists dealing with issues of politics, economics, magic and science all interwoven in the streets of the immense metropolis of New Crobuzon[65], on the suspended walkways of the city of Armada[66] and aboard the carriages of the Perpetual Train[67].

There you will find the *Remade*, reconstructed individuals who have been altered by both technology and magic into a grotesque mockery of their former selves. Although occasionally the result of a free choice, often they are the result of the Bas-Lag attitude that the punishment should fit the crime. Government scientists replace or remove body parts altogether and reshape torsos with additional constructions. Forced to find employment which suits their hybrid forms they are relegated to the fringes of society as criminals, prostitutes and dockworkers. Thus someone who would not break confidence to the city authorities has their reluctant orifice remade:

> Joshua's... remaking had been very small and very cruel. A failed burglar, he had refused to testify against his gang, and the magister had ordered his silence made permanent: he had had his mouth taken away, sealed with a seamless stretch of flesh. Rather than live on tubes of soup pushed through his nose, Joshua had sliced himself a new mouth, but the pain had made him tremble, and it was a ragged, torn, unfinished-looking thing, a flaccid wound[68].

These scientists who remake those who come under the scrutiny of the state justify their activities by reference to a higher authority, that of scientific progress. A familiar theme, the idea that progress is an inevitable force seems at times to infect the scientific worldview. Yet the Frankenstein century has questioned the very nature of that progress.

The trouble with transgenics

Nowhere is this more demonstrable than in the issue of genetically modified crops. Proponents argue that genetically modified (GM) foods present a means to rapidly progress crop characteristics including yield, pest resistance or herbicide tolerance – results which traditional methods cannot match. Furthermore, GM crops can be manipulated to fabricate completely new substances, from plastics to vaccines.

Yet the public is concerned that genetic modification techniques could result in human health, environmental and economic problems. Such was the magnitude of the debate that the British Government funded a wide-ranging public consultation exercise to survey public opinion and help drive forward public policy[69], only to reject its findings when it did not tie in with the policy directions they wished to pursue. With unanticipated responses to these new substances, inadvertent ecological toxicity and the increasing control of agriculture by both government and biotechnology corporations all a focus for disquiet, the report codified community concerns. The issue of 'Frankenstein foods' has become a locus for the worries of the public in the Frankenstein century.

Science fiction has reflected contemporary concerns about organisms which have a foreign gene (a transgene) incorporated into its genome. They have done this by transferring the dilemmas to the more cinematically appealing notion of more malevolent transgenic organisms. Although the idea of killer vegetables has been covered by science fiction[70], it was transgenic cockroaches which overran the subways in *Mimic* (1997), a transgenic spider which gave Peter Parker the powers to become *Spiderman* (2002), and a wide variety of transgenic animal DNA which allowed the government to breed supersoldiers in the television series *Dark Angel* (2002). Science drove forward advances in genetic modification, which in turn drove forward public concerns about the unalloyed application of that science; this in turn fed science fiction with sources for the exposition of plots about science escaping the test tube, which in turn helps to create the climate of distrust and suspicion.

At the heart of these concerns is the idea that, once ingested, these organisms will mutate us from the inside out. Who then do we trust? If our neighbours still look the same and act the same, how are we to know if they are infected? The paranoia and conspiracy upon which this is predicated has long antecedents within both science and science fiction. The transmission of virulent disease has embedded itself within the consciousness of the modern mind, not least because of the influenza pandemic of 1918–19, in which between fifty and one hundred million people worldwide were killed in what has been described as 'the greatest medical holocaust in history'[71]. Pandemic outbreaks loom large in the consciousness of the Frankenstein century. The 2003 SARS outbreak and the threat of Avian influenza combine to make us worried about our neighbours and friends. They become our plague vessels, not just spreading disease, but also mistrust.

Cylon, farewell, auf wiedersehen, goodbye

The idea that danger is hiding in plain sight and that our neighbours are not what they seem is a central concept in the Frankenstein century's re-imagined *Battlestar Galactica*. Based on the 1978 classic series, the reimagined *Galactica* broadly followed its predecessor, with notable twists.

Both versions maintain that in a far-off part of the Universe, twelve colonies of humans are attacked by a cybernetic race, the Cylons. Assisted by a human collaborator, Baltar (in the re-imagining a classic example of the mad scientist), the Cylons wipe out most of the populace leaving a small remnant adrift in space in a ragtag fleet led by a warship, the eponymous *Galactica*. Guided by its Commander, William Adama, the survivors' only hope is to find refuge in the fabled lost colony of Earth. The differences between the two versions lie in the portrayal of the Cylons. In the original these are large mirrored menacing androids, whose principal tactic is to demolish any opposition with brute force and ignorance. In the revised version the Cylons become more subtle, deploying a heady mixture of extreme violence and subversive cunning.

In this they are aided by their two principal incarnations. From the very beginning it is clear that the revised Cylon Centurion is a largely robotic close combat warrior. But we are also introduced to a second type of Cylon. Virtually indistinguishable from humans, these look, sound and think like us. Indeed they sometimes think they *are* us – they have become the perfect sleeper agents, well suited for acts of terrorist sabotage and able to sow panic and confusion wherever they travel.

During the course of the show it is revealed that these Cylons are themselves manufactured on a genetic level in resurrection ships. Once discovered and destroyed they are able to be re-sleeved in a new body – a body which is obviously technological on some level, as evidenced in the interaction between the Cylon agents and the technology that surrounds them. They have become, through genetic manipulation, a hybrid of man and machine. Such is the authenticity of these new Cylons that it is revealed that they can successfully mate with humans and bear offspring.

A superb piece of science fiction, its many levels of meaning and high production values revitalized the popular televisual science fiction of the early twenty-first century. Shots of hundreds of identical humanoid Cylons stretching off into the distance referenced the cloning question which has come to be part of both the scientific and science fictional landscape of the Frankenstein century.

Clone wars

The boundaries between science fiction and DNA bleed through from their common terminology. Sharing a common vocabulary, both speak of cutting, in visual science fiction, as an edit – combining two shots as opposed to the physical act of separating matter. Similarly, both splice, although genetic engineers splice

DNA not film. Actors can be cloned through visual effects so that long dead icons can be seen to interact with their more modern contemporaries; 'Cloning is popularly perceived as a way of transcending death, like being immortalized upon the silver screen'[72]. The science fiction of the twenty-first century was given a huge shot in the arm by the appearance of one of the largest stars of stage and screen in an ovine incarnation. As already mentioned at the beginning of the chapter, Dolly the sheep was the product of scientific research by Ian Wilmot and Keith Campbell at the Roslin Institute[73]. Her successful cloning resulted in the widespread application of the technology to other mammals. From early on[74] problems came to be identified which made the possible application to the human genotype unlikely. This did not stop a rush of concern in the public imagination, concerns which H. G. Wells anticipated in 1896 when his Moreau remarked:

> I began with a sheep, and killed it after a day and a half... I took another sheep, and made a thing of pain and fear and left it bound up to heal. It looked quite human to me when I had finished it; but when I went to it I was discontented... These fear-haunted, pain-driven things... they are no good for man making[75].

Such was the ignition of the public's imagination following the Roslin announcement that *Time* magazine carried an article exploring the future for cloning. That article set out four science fiction scenarios which challenged its readership. The public were asked to decide about the probity of continuing research into cloning, a process which could perhaps result in human cloning. They did this by considering the average attendance of a 'a busy morning in the cloning laboratory of the big-city hospital'[76]. Stumbling across the threshold come '(1) an aging male despot; (2) a terminally ill physics laureate; (3) a male industrialist who wants his clone to serve as his son; (4) parents who want to clone their [terminally ill] six-year-old daughter'[77]. It was only by using science fiction that *Time* felt able to sufficiently engage with the public in terms of the ethical consequences of unabashed scientific advances within the field of cloning. The public felt familiar with such explorations within science fiction: it provided a safe environment in which different futures could be extrapolated. From the nightmarish (the aging male despot) and self-serving (the industrialist's desire for an heir) to the sentimental (the terminally ill child) and scientific (physics laureate), science fiction gives us the freedom to elevate our sacred cows and despatch them without a hint of reticence, like lambs to the slaughter. The populist science fiction of the Frankenstein century reflects the concerns of the public and of science.

Nowhere is this more obvious than in the newest incarnations of the twentieth century's most popular science fictional product, *Star Wars*. The Frankenstein century's incarnation deployed waves of clones in its twenty-first century prequels. The overall story of the three George Lucas-helmed blockbuster hits deals with the development and utilization of these vat-bred warriors, the second part of the trilogy even going so far as to be named *Star Wars Episode II: Attack of the Clones*. Developed in secret by a race of corporate scientists, these clone warriors become tools of an oppressive dictatorship bent on universal domination. In the trilogy it is the hero Obi Wan Kenobi, a Jedi knight played by Ewan McGregor, who fights against the conspiracy.

Ewan McGregor stars again in a cloning story, this time as a clone himself, in the 2005 blockbuster *The Island*. A clone of various precursor tales, this science fiction film liberally borrows from, amongst others, Huxley's *Brave New World*, the 1979 low-budget science fiction schlocker *Parts: The Clonus Horror*, *Logan's Run* (1976), George Lucas's *THX 1138* (1971) and *Blade Runner* (1982), but most closely seems to mirror the 1996 Michael Marshall Smith novel *Spares*. Again a profit motivated group of scientists are the antagonists. This time they have cloned various wealthy individuals in order to be able to harvest the clones' organs should their clients ever require a transplant. *The Island* is noted not just for the relatively low impact it made at the box office but also for the level of corporate sponsorship and product placement that the film boasted. With products from MSN Search, Xbox, Puma, Reebok, NBC, NFL, Budweiser, Apple Computer, Aquafina, General Motors, DaimlerChrysler, Mack, Coca-Cola, Speedo, TAG Heuer, Amtrak, Ben & Jerry's and Nokia all featured, the film is at times reduced to an extended advertising campaign.

The corporate presence in cloning represented in the science fictional blockbusters of the Frankenstein century has its real-life parallels in the biogenetics industry. Nowhere is the clash of private industry versus public investment more demonstrable than in the Frankenstein century's attempts to sequence the human genome itself.

Although whole genome mapping had existed since 1972[78] the decision by the US Government to try to clone the human genome was made in 1987 when it was asserted that 'Knowledge of the human genome is as necessary to the continuing progress of medicine and other health sciences as knowledge of human anatomy has been for the present state of medicine'[79]. The three billion dollar project was conducted by an international consortium led by the US with contributors from China, France, Germany, Japan and the United Kingdom. However, this public project had a corporate rival.

Celera Genomics, whose leading figures had led the first successfully sequencing of an entire organism's genome, were determined to sequence the human genome faster and more cheaply than the public project. Partially this was possible because they chose to use the quicker, yet less robust, shotgun method of sequencing, in which the sample is blasted into small pieces, as opposed to the slower and more stable clone-by-clone method used by the public project. Celera's success, at one tenth of the cost of the public project, was tainted, however, when they published their results in 2001 – results which they had fused with the available data from the public project so as to render the size of their own contribution indistinguishable from the whole. Claims that the use of the shotgun approach accelerated the public project[80] have been balanced by accusations that Celera's attempts to withhold data from the public arena for material gain retarded scientific progress as a whole[81].

There is no gene for the human spirit[82] – Gattaca

The commercialization of genetic information found its science fictional expression in the 1997 film *Gattaca*. The title is a hybrid term recalling both one of America's most famous prisons, Attica, and the abbreviations for the DNA nucleotide bases Adenine, Thymine, Cytosine and Guanine. Presenting a nightmarish vision of a society driven by new eugenics, the offspring of all but the lowest classes are genetically engineered from conception to be the best they can be. The reprogenics outlined in *Gattaca* allow individual parents to choose whether to interfere with their child's development or not. Based on a model developed by Professor Lee M. Silver at Princeton University[83], reprogenics merges reproductive and genetic technologies to allow parental choice in the production of a designer child. Although applications include eliminating inheritable diseases, Silver's work assumes that very quickly market forces will assert themselves. This will in turn trigger massive social changes, the consequences of which would be to create a two-tiered society. Bio-ethicist James Hughes asserts that this new biopolitics is intrinsically oppositional, the two sides being:

> transhumanists (the progressives) and, at the other end, the bio-Luddites or bio-fundamentalists. Transhumanists welcome the new biotechnologies, and the choices and challenges they offer, believing the benefits can outweigh the costs. In particular, they believe that human beings can and should take control of their own biological destiny, individually and collectively enhancing our abilities and expanding the diversity of intelligent life. Bio-fundamentalists, however,

reject genetic choice technologies and 'designer babies,' 'unnatural' extensions of the life span, genetically modified animals and food, and other forms of hubristic violations of the natural order. While transhumanists assert that all intelligent 'persons' are deserving of rights, whether they are human or not, the biofundamentalists insist that only 'humanness,' the possession of human DNA and a beating heart, is a marker of citizenship and rights[84].

The *Star Trek* spin-off *Deep Space Nine* dealt with the genetically modified humans by making such actions illegal in the Federation, forcing the space station doctor, Julian Bashir, to hide his ancestry[85]. *Gattaca*, however, takes the opposite view. In *Deep Space Nine* it is Silver's GeneRich who are persecuted and imprisoned, whilst in *Gattaca* it is Silver's GenePoor who find themselves discriminated against.

Society's system of control in *Gattaca* is via biometric scans, similar to those now adopted by governments around the world for passport control. This allows for the illegal yet commonplace discrimination between the 'valids', those who were modified, and the 'in-valids', or those born by faith-births. Industry and commerce staff their projects with the designer ideal and allow those who are the product of natural selection to fester in menial jobs.

Such is the case for *Gattaca's* protagonist Vincent. As a natural birth his defects preclude his entry into the space program, which is his dream. He therefore fakes his identity, taking instead the life history of a GeneRich but paraplegic valid, Jerome Eugene Morrow. As the plot unfolds, Vincent, desperate to retain his assumed identity, is embroiled in a murder plot, at the crime scene of which his assumed DNA is present. At the same time he is pursuing a romantic liaison with a co-worker, Irene Cassini. The film climaxes with Vincent blasting off into space as Jerome commits suicide, calling into question the very basis of reprogenic selection. Both successfully escape from the biological prison that they were fated to endure, Vincent by escaping Earth itself and Jerome by abandoning the disabled body which confined him.

The superficial nature of judgements based upon the disputation of a person's DNA is reinforced by the director Andrew Niccol's decision to use fashion models as extras and shoot the film with an overtly stylish veneer. Niccol, who also directed *S1m0ne* (mentioned in the Computer Age), has a sister who suffers from multiple sclerosis, and it is perhaps this fact which leads to the inclusion on the DVD of a scene deleted from the film's final cut but shown as an optional extra.

The sequence displays a mosaic of famous people who might not have been born if a reprogenic policy had been existence. These include Abraham Lincoln (Marfan's syndrome), Emily Dickinson (manic depression), Vincent van Gogh (epilepsy), Albert Einstein (dyslexia), John F. Kennedy (Addison's disease), Rita Hayworth (Alzheimer's disease), Ray Charles (primary glaucoma), Stephen Hawking (amyotrophic lateral sclerosis), and Jackie Joyner-Kersee (asthma). The sequence ends with the words: 'Of course, the other birth that may never have taken place is your own'[86].

A science fictional world

Writing on his blog, itself the product of a science fictional vision wrought amidst the Computer Age, the acclaimed Frankenstein century author and graphic artist Warren Ellis commented that 'we live in a science fictional world. Not the one everyone expected, of course – no jetpacks. But good science fiction, challenging science fiction, is never about the future we expect. SF has never been about predicting the future. It's been about laying out a roadmap of possibilities, one dark street at a time, and applying that direction to the present condition'[87]. The warning that it is 'one dark street at a time' is followed up in Ellis's own LiveJournal comments:

the FDA has approved cloned meat for human consumption, and also decided that packaging information need not include the provenance of the meat. Right now, of course, cloned meat is too expensive to take to market. But, in years to come, it's entirely conceivable that your burgers will come from cloned animals. What's the big deal with that, you ask? Well, check out Dolly the sheep. You can't, can you? That's because the science-sheep that nobody dreamt of died young, bones crumbling and infection-mangled organs falling out... that's the science-fictional world you're living in today[88].

Ellis's dismay is echoed in the words and thoughts articulated some ten years earlier by the father of Cyberpunk, William Gibson, who commented in 1997 'you're living in an overlapping batch of science-fiction scenarios. AIDS is a science fiction scenario. The decay of the ozone layer is a science fiction scenario'[89]. If we sci-ence fictionally travel back through time, touching down here and there in our own narrative we find that this conception of living at the edge, of being part of a brave new world has remained ever present. That such a world alternates between dystopian visions of science run amok and of societies saved and elevated by scientific endeavour is the stuff that stories are made of.

'... Where no one has gone before'

In the aftermath of the launches of Sputnik 1 and 2, a Presidential Science Advisory Committee (PSAC), was formed to assess the appropriate direction and pace for the US space program. The PSAC focused heavily on the scientific aspects of the space pro-gram. With President Dwight D. Eisenhower's endorsement, on 26 March 1958, it released a report outlining the importance of space activities, but recommended a cautiously measured pace. Eisen-hower's preface remarked that 'This is not science fiction. This is a sober, realistic presentation prepared by leading scientists'[90].

Directly following on from Eisenhower's introduction, Dr James Killian, Science Advisor to the President, remarked that in the proposal for a national space program:

It is useful to distinguish among four factors which give importance, urgency, and inevitability to the advancement of space technology. The first of these factors is the compelling urge of man to explore and to discover, the thrust of curiosity that leads men to try to go where no one has gone before[91].

You may recognize that final portion of rhetoric. It made its way into the opening narration of the most influential science fictional television franchise to date. Years later the first vessel in a pioneering new approach to spaceflight from that same administration took as its name the title of the fictional ship within that series. The first Space Shuttle, *Enterprise*, acknowledged the debt that NASA owed to science fiction in inspiring in a myriad ways its own evolution.

Science drove fiction and that fiction in turn drove science. They may have different engines but both propel us forward.

To claim that there is an element of prophecy incumbent within that science is to misunderstand its intention. As one of the key participants of the American space race, Neil Armstrong, accurately commented 'Science has not yet mastered prophecy. We predict too much for the next year and yet far too little for the next ten'[92]. Science fiction likewise does not prophesy the future; it does, however, lay out possibilities. Science fiction logically extrapolates analytical scenarios from the crucible of science, but that is not its task. Its task is to inspire those reading to open their minds up to the inherent possibilities that exist within science; not to be constrained by reductionist thinking that dwells on the here and now, but to allow our minds to soar through a Universe of wonder and imagination. Science fiction helps to train our intellects to accept our imagination as a useful tool within science's toolbox. Its mode of thinking reorientates our perspective so that, just as the new world explorers of the Age of Discovery saw new horizons, so too will we. As Darwin revolutionized science in the Mechanical Age by utilizing a new perspective, so too can we. As the fictional authors of the Atomic Age extrapolated the dangers of scientific developments, so too can we.

The science fiction of the new millennium will continue to be driven by science. The Frankenstein century's extrapolations of our own evolutionary and genetic future will continue to remain sceptical of the promises of scientific materialism. It will continue to reflect the questions and concerns of the wider public's imagination. It will continue to be driven by science.

Science too will continue to be urged onward through science fiction's visionary situations. It will continue to identify hitherto unanticipated areas of exploration through its fiction. It will continue to boldly go where no one has gone before.

References

1. Malik, K. (2000) *Man, Beast and Zombie*. Weidenfeld and Nicolson, London.
2. Sagan, C. (1994) *Pale Blue Dot*. Random House, New York.
3. Vonnegut, K. (1969) *Slaughterhouse 5*. Delacorte Press, New York, p. 73.
4. *Ibid.*, p. 73.
5. Rose, M. (1982) *Alien Encounters: Anatomy of Science Fiction*. Harvard University Press, Cambridge, MA, p. 32.
6. *Ibid.*, p. 32.
7. *Ibid.*
8. *Ibid.*
9. Rose, M. (1982) *Alien Encounters: Anatomy of Science Fiction*. Harvard University Press, Cambridge, MA, p. 38.
10. Aldiss, B. (1973) *Billion Year Spree*. Weidenfeld and Nicolson, London.
11. Isaacs, L. (1977) *Darwin to Double Helix: The Biological Theme in Science Fiction*. Butterworth, London, p. 11.
12. *Ibid.*, p. 11.
13. Hoskins, M. (1997) *The Cambridge Illustrated History of Astronomy*. Cambridge University Press, Cambridge.
14. Sagan, C. and Shklovskii, I. S. (1966) *Intelligent Life in the Universe*. Holden-Day, San Francisco.
15. Dick, S. J. (1996) *The Biological Universe*. Cambridge University Press, Cambridge.

16. Isaacs, L. (1977) *Darwin to Double Helix: The Biological Theme in Science Fiction*. Butterworth, London, p. 6.
17. Henkin, L. J. (1963) *Darwinism in the English Novel 1860–1910*. Corporate Press, New York.
18. Isaacs, L. (1977) *Darwin to Double Helix: The Biological Theme in Science Fiction*. Butterworth, London, p. 19.
19. Aldiss, B. (1973) *Billion Year Spree*. Weidenfeld and Nicolson, London, p. 133.
20. Stapledon, O. (1999) *Last and First Men*. Millennium, London, p. xiii.
21. *Ibid.*, pp. xv–xvi.
22. *Ibid.*, p. xiii.
23. Isaacs, L. (1977) *Darwin to Double Helix: the Biological Theme in Science Fiction*. Butterworth, London, p. 24.
24. Stapledon, O. (1999) *Last and First Men*. Millennium, London, Chapter IV, 4.
25. *Ibid.*
26. Quoted in Isaacs, L. (1977) *Darwin to Double Helix: the Biological Theme in Science Fiction*. Butterworth, London, p. 47.
27. Stapledon, O. (2001) *Star Maker*. Millennium, London, p. 13.
28. Dick, S. J. (1993) *Consequences of Success in SETI: Lessons from the History of Science*, Vol. 74. Progress in the Search for Extraterrestrial Life, 1993 Bioastronomy Symposium, Conference Proceedings of the Astronomical Society of the Pacific, pp. 521–32.
29. Shapley, H. (1958) *Of Stars and Men: Human Response to an Expanding Universe*. Elek Books, London.
30. Struve, O. (1961) *The Universe*. MIT Press, Cambridge, MA.
31. Berenzden, R. (1975) *Copernicus Yesterday and Today* (Vistas in Astronomy 17) (eds. A. Beer and K. Strand). Pergamon Press, New York, pp. 65–83.
32. Stapledon, O. (2001) *Star Maker*. Millennium, London, p. 89.
33. Clarke, A. C. (1956) *The City and the Stars*. Harcourt, Brace, New York, pp. 174–5.
34. Clarke, A. C. (1972) *Report on Planet Three & Other Speculations*. Harper & Row, New York, p. 89.
35. Clarke, A. C. (1990) *Childhood's End*. Harcourt, Brace, New York, p. 8.
36. *Ibid.*, p. 8.
37. Dick, S. J. (1996) *The Biological Universe*. Cambridge University Press, Cambridge.
38. Youngblood, G. (1970) *Expanded Cinema*. Dutton, New York, p. 139.

39. Kaufmann, W. (1982) *The Portable Nietzsche*. Viking, New York, p. 124.
40. *Ibid.*, p. 227.
41. Rose, M. (1982) *Alien Encounters: Anatomy of Science Fiction*. Harvard University Press, Cambridge, MA, p. 32.
42. *Ibid.*, p. 144.
43. Clarke, A. C. (1984) *1984: A Spring of Futures*. Ballantyne, New York, p. 111.
44. Miller, M. C. (1994) 2001: a cold descent. *Sight and Sound*, **4**(1), 24.
45. Burns, J. F. (1997) For Arthur C. Clarke, what is paradise without praise? *New York Times*, 1 April.
46. LoBrutto, V. (1977) *Stanley Kubrick: A Biography*. Penguin, London.
47. Clarke, A. C. (1972) *2001: A Space Odyssey*. Hutchinson, London, p. 7.
48. Pilgrim, I. (1984) The Too-Good-to-be-True Paradox and Gregor Mendel. *Journal of Heredity*, #75, pp. 501–2.
49. Edelson, E. (2001) *Gregor Mendel and the Roots of Genetics*. Oxford University Press, London.
50. Fisher, R. A. (1918) The correlation between relatives on the supposition of Mendelian inheritance. *Philosophical Transactions of the Royal Society of Edinburgh*, **52**, 399–433.
51. Dawkins, R. (1996) *River Out of Eden*. Phoenix, London, pp. 43–4.
52. Shakespeare, W. (1611) *The Tempest*, Act V, Scene I.
53. Huxley, A. (1977) *Brave New World*. Triad Panther, London, p. 15.
54. *Ibid.*, p. 33.
55. *Ibid.*, p. 45.
56. Donne, J. (1633) *Meditation XVII*.
57. Huxley, A. (1977) Foreword to *Brave New World*. Grafton, London, p. 9.
58. Huxley, A. (1962) *Island*. Chatto & Windus, London.
59. Campbell, D. (2002) Island of dreams. *The Guardian*, 27 April.
60. Jackson, M. (1986) *Life Story – Race for the Double Helix*. A&E Television Networks.
61. Frayling, C. (2005) *Mad, Bad and Dangerous?* Reaktion Books, London, p. 105.
62. Lam, C. *Personal Diary – Recollection of Byron*.
63. Sagan, C. (1995) *The Demon Haunted World*. Random House, New York, p. 245.
64. Miéville, C. (2003) Guest editorial. In *The Third Alternative 35*, Summer.

65. Miéville, C. (2000) *Perdido Street Station*. PanMacmillan, London.
66. Miéville, C. (2002) *The Scar*. PanMacmillan, London.
67. Miéville, C. (2004) *Iron Council*. PanMacmillan, London.
68. Miéville, C. (2000) *Perdido Street Station*. PanMacmillan, London, p. 31.
69. Department of Trade and Industry (2003) *GM Nation? The Findings of the Public Debate*.
70. *Attack of the Killer Tomatoes!* (1978) Dir. John De Bello.
71. Potter, C. W. (2006) A history of influenza. *Journal of Applied Microbiology*, **91**(4), 572–9.
72. Nottingham, S. (1999) *Screening DNA – Exploring the Cinema*. DNA Books, p. 1.
73. Campbell, K. H. S., McWhir, J., Ritchie, W. A. and Wilmut, A. (1996) Sheep cloned by nuclear transfer from a cultured cell line. *Nature*, **380** (6569), 64–6.
74. Shiels, P.G., *et al.* (1999) Analysis of telomere lengths in cloned sheep. *Nature*, **399** (6734), 316–17
75. Wells, H. G. (1896) *The Island of Dr Moreau*. BCA, London, p. 135.
76. Kluger, J. (1997) Will we follow the sheep? *Time*, 10 March.
77. Barr, M. (2000) 'We're at the start of a new ball game and that's why we're all real nervous': Or, Cloning – Technological Cognition Reflects Estrangement from Women in Learning from Other Worlds (ed. P. Parrinder). Liverpool University Press, Liverpool, p. 201.
78. Min Jou, W., Haegeman, G., Ysebaert, M. and Fiers, W. (1972) Nucleotide sequence of the gene coding for the bacteriophage MS2 coat protein. *Nature*, **237**, 82–8.
79. Human Genome Program, U.S. Department of Energy, Human Genome News (v1n1).
80. Shreeve, J. (2004) *The Genome War: How Craig Venter Tried to Capture the Code of Life and Save the World*. Ballantine, New York.
81. Sulston, J. and Gerry, G. (2002) *The Common Thread*. Bantam Press, New York.
82. *Gattaca* (1997) dir. Niccol.
83. Silver, L. M. (1998) *Remaking Eden: Cloning and Beyond in a Brave New World*. Harper Perennial, New York.
84. Hughes, J. (2002) Democratic transhumanism. *Transhumanity*, 28 April.
85. Doctor Bashir, I Presume? *Star Trek: Deep Space Nine*. Paramount, 1997.
86. *Gattaca* (1997) dir. Niccol.
87. Ellis, W. (2006) *Flying Frogs And Crashed Rocketships*. http:// www.warrenellis.com/, 29 December.

88. Ellis, W. (2007) `http://www.warren-ellis.live-journal.com/77339.html`, 2 January.

89. Gibson, W. (2000) quoted in Barr, Marlene: 'We're at the start of a new ball game and that's why we're all real nervous': Or, Cloning-Technological Cognition Reflects Estrangement from Women in Learning from Other Worlds (ed. P. Parrinder). Liverpool University Press, Liverpool, p. 194.

90. Eisenhower, D. D. Preface to President's Science Advisory Committee Introduction to Outer Space, 26 March 1958, p. 1. NASA Historical Reference Collection, NASA History Division, NASA Headquarters, Washington, DC.

91. President's Science Advisory Committee Introduction to Outer Space 26 March 1958, p. 1. NASA Historical Reference Collection, NASA History Division, NASA Headquarters, Washington, DC.

92. Armstrong, N. (source unknown).

INDEX

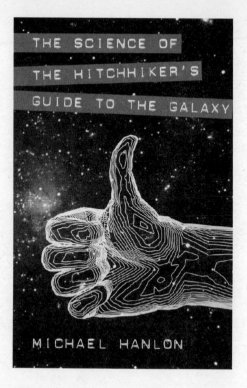

THE SCIENCE OF THE HITCHHIKER'S GUIDE TO THE GALAXY
by Michael Hanlon

MACMILLAN; ISBN 1–4039–4577–2; £16.99/$24.95; HARDCOVER
ISBN: 0–230–00890–9; £8.99/$14.95; PAPERBACK

"Adopting Adams' witty, punchy style, Hanlon's guide is a fun and vivid read. The science twinkles a little more than usual in such a zany setting... he tackles a wide range of cutting-edge topics with depth and authority." *Nature*

"Hanlon's book probes the possibilities inside the fiction with wit and scientist humour – not that you have to be a boffin to enjoy these ruminations, merely curious, as the late Adams himself clearly was." *The Herald*